T0181746

Studies in Computational Intelligence

Volume 572

Series editor

Janusz Kacprzyk, Polish Academy of Sciences, Warsaw, Poland
e-mail: kacprzyk@ibspan.waw.pl

About this Series

The series "Studies in Computational Intelligence" (SCI) publishes new developments and advances in the various areas of computational intelligence—quickly and with a high quality. The intent is to cover the theory, applications, and design methods of computational intelligence, as embedded in the fields of engineering, computer science, physics and life sciences, as well as the methodologies behind them. The series contains monographs, lecture notes and edited volumes in computational intelligence spanning the areas of neural networks, connectionist systems, genetic algorithms, evolutionary computation, artificial intelligence, cellular automata, self-organizing systems, soft computing, fuzzy systems, and hybrid intelligent systems. Of particular value to both the contributors and the readership are the short publication timeframe and the worldwide distribution, which enable both wide and rapid dissemination of research output.

More information about this series at http://www.springer.com/series/7092

David Camacho · Sang-Wook Kim
Bogdan Trawiński

Editors

New Trends in Computational Collective Intelligence

 Springer

Editors
David Camacho
Autonomous University of Madrid
Madrid
Spain

Bogdan Trawiński
Wrocław University of Technology
Wroclaw
Poland

Sang-Wook Kim
Hanyang University
Seoul, Korea
Republic of (South Korea)

ISSN 1860-949X   ISSN 1860-9503   (electronic)
ISBN 978-3-319-36489-6 ISBN 978-3-319-10774-5   (eBook)
DOI 10.1007/978-3-319-10774-5

Springer Cham Heidelberg New York Dordrecht London

Printed on acid-free paper

Springer is part of Springer Science+Business Media (www.springer.com)

Preface

Computational Collective Intelligence methods and algorithms are one the current trending research topics from areas related to Artificial Intelligence, Soft Computing, and Data Mining among others. Computational Collective Intelligence (CCI) is a rapidly growing field that is most often understood as an AI sub-field dealing with soft computing methods which enable making group decisions or processing knowledge among autonomous units acting in distributed environments. Web-based Systems, Social Networks, and Multi-Agent Systems very often need these tools for working out consistent knowledge states, resolving conflicts, and making decisions.

These techniques and methods use the behavior and interactions from a set of "autonomous entities" to build a new kind of intelligence that emerges from these interactions. The massive interactions between these intelligent entities comprise several activities including collaboration, competition, opinion expressing, value exchange, and message exchange. These computational methods are usually classified as "bioinspired" and some of the methods and techniques used in this area have been adapted from sociology, biology, economy or political sciences. These interactions can involve for example consensus reaching, decision making, social choice or other means for quantifying mass activities.

The chapters included in this volume cover a selection of topics and new trends in several domains related to Collective Computational Intelligence: Language and Knowledge Processing, Data Mining Methods and Applications, Computer Vision, and Intelligent Computational Methods. The book consists of 20 extended and revised chapters based on original works presented during a poster session organized within the 6th International Conference on Computational Collective Intelligence (ICCCI 2014) that was held between 23 and 26 of September 2014 in Seoul, Korea. The book is divided into four parts. The first part is entitled "Language and Knowledge Processing" and consists of six chapters that concentrate on many problems related to NLP and CCI problems, including: text mining, quality of collective knowledge, and translation quality. The second part of the book is entitled "Data Mining Methods and Applications" and encompasses six chapters. The authors present applications related to several intelligent computational methods like clustering and collaborative filtering in areas like prediction in telecommunication networks, sensor networks, and human-face recognition.

third part of the book is entitled "Computer Vision Techniques" and contains five chapters devoted to processing methods for computer vision and knowledge management in applications such as: game storytelling, mobile applications, tourism, and Google Street view. The last part of the book, entitled "Intelligent Computational Methods" comprises three chapters related to the application of CCI methods in new application domains, such as: videogames, personalized social search, and personalized travel scheduling domains.

The editors believe that this book can be useful for graduate and PhD students in computer science as well as for mature academics, researchers and practitioners interested in the methods and applications of collective computational intelligence in order to create new intelligent systems.

We wish to express our great attitude to Prof. Janusz Kacprzyk, the editor of these series, and to Dr. Thomas Ditzinger from Springer for their interest and support for our project.

The last but not least we wish to express our great attitude to all authors who contributed to the content of this volume, and to reviewers that provide us the necessary support to improve the quality of published papers.

David Camacho
Sang-Wook Kim
Bogdan Trawiński

Contents

Part III: Computer Vision Techniques

Part IV: Intelligent Computational Methods

Part I
Language and Knowledge Processing

An Analysis of Influence of Consistency Degree on Quality of Collective Knowledge Using Binary Vector Structure

Marcin Gębala, Van Du Nguyen, and Ngoc Thanh Nguyen

Institute of Informatics, Wroclaw University of Technology,
Wyb.St. Wyspianskiego 27, 50-370 Wroclaw, Poland
maarcin.gebala@gmail.com,
{van.du.nguyen,Ngoc-Thanh.Nguyen}@pwr.edu.pl

Abstract. As we know, determining the knowledge of a collective is an important task. However, there exists another important issue with its quality. The quality reflects how good the collective knowledge is. It is useful in some cases such as: to add or remove some knowledge members to improve quality of collective knowledge or evaluate whether collective knowledge is good enough or not. In this paper, we consider consistency functions that proposed by taking into account both density and coherence factors. Then we analyze influence of their values on the quality of collective knowledge using binary vector structure. The experiments showed that both density and coherence have a significant influence on the quality of collective knowledge.

Keywords: collective knowledge, consensus theory, consistency functions.

1 Introduction

Up to now, the idea of processing collective knowledge has been controversial [11]. There exist many conceptions for processing collective knowledge such as: justified true belief or acceptance held or arriving by groups as plural subjects [12]; the sum of shared contributions among community members [10]. In this paper we assume collective knowledge is the common state of knowledge of a collective as a whole [9]. That could be considered as the representative of members in a collective. For each collective, besides the problem determining the knowledge of collective also called collective knowledge. There exists another important issue with its quality. That is whether collective knowledge is good enough or not. We mention to evaluating the quality of collective knowledge. It is useful in some cases such as: to add or remove some knowledge members to improve quality of collective knowledge or evaluate how good the collective knowledge is. In [7], author presented that the factors such as: density and coherence of members in a collective have influence on the quality of collective knowledge. The factors mean how close or crowed collective members are? They were also taken into account in proposing consistency functions to measure consistency degree of each collective. However, author hasn't analyzed relationship

D. Camacho et al. (eds.), *New Trends in Computational Collective Intelligence*,
Studies in Computational Intelligence 572, DOI: 10.1007/978-3-319-10774-5_1

between consistency degree and the quality of collective knowledge. In [8], author proposed another aspect of quality. That is, distance from collective knowledge to real knowledge state with assumption each collective member presents some degree of real knowledge state. The better collective knowledge is the closer knowledge state to the real knowledge state.

In this paper, we analyze influence of consistency degree on quality of collective knowledge by using binary vector structure. This is a simple structure and useful for many IT applications such as data transaction in market basket data, document clustering [4].With each collective, we need to determine consistency values, evaluate its quality, and then analyze relationship between consistency values and the quality. For this aim, some basic notions about collective of knowledge states, consistency functions, and algorithm to determine collective knowledge are presented. Then we analyze experiments to determine how consistency values influence on the quality of collective knowledge. In other words, we try to prove the hypothesis "*The higher consistency, the better is quality of collective knowledge?*" That is the main problem what we investigate in this paper. This approach is missing in the literature.

The remaining parts of this paper are organized as follows: section 2 presents the structure of binary vector, collective of knowledge states, collective knowledge, and consistency functions. In section 3, problem determining representative of a collective is presented. In this section, we introduce consensus methodology, algorithm for consensus choice. In section 4, we present experiments with different collective size and different length of binary vector. Then we analyze the results in section 3 and examine relationships between consistency degree and the quality of collective knowledge. Some conclusions and future works are presented in section 5.

2   Basic Notions

2.1   Collective of Knowledge States

In this paper, we assume each knowledge state is a binary vector in which represented an opinion of an agent about a subject or matter in the real world. Each opinion contains a number of attributes; the binary value 1 indicates the present of an attribute, otherwise value 0. Therefore, U is a finite set of binary vectors (Universal set) with the given binary vector length. *For example, with binary vectors of length 10, set U has 1024 (2^{10}) elements.* Collective is a set (with repetition) of knowledge states.

Definition 1. *The structure of a collective is defined as follows:*

$$X = \{x_1, x_2, \dots, x_n\}$$

where x_i is a binary vector; n is number of collective members. All binary vectors in a collective are in the same length.

2.2 Collective Knowledge

Collective knowledge is understood as a representative of each collective. It is an element from set U that satisfies some criteria. *For example: the sum of distances or squared distances from collective knowledge to collective members is minimal.* In this paper, we consider the following definition:

Definition 2. Collective knowledge of a collective is a knowledge state that satisfies:

$$d(x,X)^i = \min_{y \in U} d(y,X)^i$$

where d(x,X) is sum of distances from x to elements of X, $i \in \{1,2\}$.

2.3 Quality of Collective Knowledge

The quality of collective knowledge is a quantity which reflects how good a collective knowledge is. According to [7], it depends on the ratio of the sum of distances from collective knowledge to elements of collective. We assume that collective knowledge of collective X called x. Then the quality of collective knowledge is measured by the following formula:

$$\hat{d}(x,X) = 1 - \frac{d(x,X)}{M}$$

where M is the size of collective X.

2.4 Consistency Degree of Collective

According to [7], the consistency degree presents the coherence level of collective members. It could be measured through the distances between members. Also in that work, author introduced five consistency functions to measure consistency degree of collectives. The consistency function is defined as the following definition:

Definition 3. The consistency degree of a collective is defined by c function as follows:

$$c: \Pi(U) \to [0,1]$$

where $\Pi(U)$ is set of nonempty subset with repetition of U.

However, in this work, we concentrate on analyzing influence of the functions c_2, c_3, and c_4 on the quality of collective knowledge. The remaining functions are not significantly influent or identical to the quality of collective knowledge. Besides, in what follows $d(x,y)$ is a distance function to measure different between collective members. In the context of the paper, with binary vector structures we use Manhattan function. This function will be mentioned in section 3. However, firstly, we consider some related parameters that used in these consistency functions:

- Maximal average distances from an element to the rest:

$$Diam(w^X) = \max_{x \in X} w_x$$

Where $w_x = \frac{1}{M-1}\sum_{y \in X} d(x,y)$

- The average distance of collective X:

$$d_{mean}(X) = \begin{cases} \dfrac{1}{M(M-1)}\displaystyle\sum_{x,y \in X} d(x,y) \; for M > 1 \\ 0 \hspace{4.5cm} for M = 1 \end{cases}$$

- The total average distance in collective X:

$$d_{t\_mean}(X) = \frac{\sum_{x,y \in X} d(x,y)}{M(M+1)} = \frac{(M-1)}{(M+1)} \sum_{x,y \in X} d(x,y)$$

Then, we have the following consistency functions:

$$c_2(X) = 1 - Diam(w^X)$$

$$c_3(X) = 1 - d_{mean}(X)$$

$$c_4(X) = 1 - d_{t\_mean}(X)$$

Some commentaries for these consistency functions: according to function c_2, if the maximal average distance between an element of collective X and other members of this collective is small. That is the members of collective X are near each other then the consistency should be high. This function takes into account the density and coherence of collective members. Function c_3 takes into account the average distance among members of X. The larger the average value, the smaller the consistency value and vice versa. This function could be considered as representative for consistency because it reflects a large degree the density of collectives. Function c_4 is dependent on c_3 and vice versa. It takes into account the total average distance between members of X.

3   Determining Collective Knowledge

In order to analyze relationship between consistency degree and the quality of collective knowledge of each collective, we need to calculate the consistency function values and determine the representative or collective knowledge of each collective. In this paper, we proposed an algorithm based on consensus method. This method has been proved to be useful in solving conflicts and should be also effective for knowledge inconsistency resolution and knowledge integration [6].

3.1   Consensus Methodology

A consensus choice has usually been understood as a general agreement in situations where parties have not agreed on some matter [2]. The oldest consensus model was worked out by Condorcet, Arrow and Kemeny [1]. This model is used to solve such conflicts in which the content may be represented by two well-known problems: Alternatives Ranking Problem and Committee Election Problem. By consensus choice, there are two well-known criteria: "1-Optimality" and "2-Optimality" or O_1 and O_2 for short. Criterion O_1 means the sum of distances between the collective knowledge and the collective members to be minimal whereas O_2 means the sum of the squared distances between the collective knowledge and the collective members is minimal. Although criterion O_1 is more popular than O_2 for consensus choice and some optimal tasks, criterion O_2 is better than O_1 in some situations [7].

3.2   Distance Function

In order to propose algorithms for determining collective knowledge, it is most important to introduce distance function. In the context of this paper we use Manhattan function for measuring difference between knowledge states. However, to evaluate the consistency degree we need to normalize that function. The function could be normalized by dividing its values by the length of binary vector. That is:

$$\partial(x, y) = \frac{d(x, y)}{n}$$

where $\partial(x, y) \in [0,1]$, $d(x, y) = \sum_{i=1}^{n} |x_i - y_i|$, *n is the length of binary vector.*

3.3   Algorithm for Determining Collective Knowledge

Below we present an algorithm used to determine the representative of collectives. The idea of algorithm is based on an intuition, that the opinions which are most frequent among members of the collective and should belong to the representative of the collective. To determine the representative for all members in the collective we compare the number of occurrences of values 0 and 1 in the subsequent positions. The value that is more frequent for the considered position is then added to the representative. The algorithm based on O_1 criterion, which the sum of distances between the collective knowledge and the members of collective is minimal.

Input: Collective $X \in \Pi(U)$

Output: vector x^* *which minimizes the sum of distances to members in collective X*

Procedure:

BEGIN

> *Initialize L: = length of binary vector,* x^* *is empty binary vector of length L*
> > *For i =0to L do:*
> > > *Calculate numberOfOnes, numberOfZeros at* i^{th} *all members of X*

If numberOfOnes $\geq$ numberOfZeros then
$$x^*[i] := 1$$
Else
$$x^*[i] := 0$$
Return vector x^*

END

4   Experiments and Their Analysis

4.1   Experiments

In this section, the hypothesis that mentioned in the Introduction will be proved through experimental results. Each experiment consists of consistency function values and the quality of collective knowledge. However, firstly, we consider the generation mechanism and environments for experiments. The experimental collectives were randomly generated by simulation with parameters as mentioned in Table 1. The simulation was built based on Python to generate collectives and calculate theirs consistency functions values as well as evaluate the quality of corresponding collective knowledge. In order to give objective results, we proposed test cases with different number of collective members and different length of binary vectors. This helps us sufficiently identify influence of consistency degree on the quality of collective knowledge. The processing experiments take the following steps. Firstly, we generate a collective of binary vectors with specified number of collective members and the length of binary vector. Then, we calculate the values of consistency functions as presented in section 2.2. Next, we apply the algorithm described in section 3.2 to obtain the collective knowledge and calculate its quality. The following table is number of members and collective size that we used to experiment.

Table 1. Parameters of collectives

Length	Collective size
5	10, 50, 100
10	100, 500, 1000
15	100, 500, 1000

In order to make sure of the experiments, in each set of parameters from Table 1 we repeated 25 iterations. In Table 2, we present experiments of collectives with 500 members of length 10. In each test, we present the values of consistency functions C2, C3, C4 and the quality of collective knowledge Q. However, from Table 2, we couldn't give any statement about relationship between consistency function values and the quality of collective knowledge. This is the task of the remaining parts of the paper. Through experiments analysis, we will concentrate on proving the hypothesis *"Whether the higher consistency degree, the better is the quality of collective knowledge?"* that mentioned in the Introduction. In other words, we need to analyze influence of consistency function values C2, C3, C4 on the quality Q. This is one of the most important sections of the paper.

4.2   Charting Analysis

One of the most effective techniques in presenting data is using chart. It provides us with a better understanding about the data and we can give some visual statements about relationship between variables. In this section we present charts for experiments with collectives (500, 100, 10 members) of length 10. Then in the next section, we use Spearman correlation coefficient to verify them. That also aims to analyze influence of consistency degree of collective on the quality of collective knowledge.

Table 2. Results for 500 members of length 10

No	C2	C3	C4	Q
1	0.48257	0.49988	0.50187	0.5184
2	0.48417	0.49973	0.50173	0.518
3	0.48036	0.49984	0.50184	0.5186
...	...	...	...	...
24	0.48136	0.49996	0.50195	0.5196
25	0.482365	0.49989	0.50189	0.5174

Firstly, Fig. 1 contains the results for experiments with 500 members of collective of length 10. Visually, we can see that the quality of collective knowledge is highly dependent on the C2 consistency. That is, the higher the values of C2 consistency, the lower the quality of determined collective knowledge and vice versa. From this fact, we can infer that the more different the opinions of collective members, the better the credibility of collective knowledge. This interpretation is coherent with the conclusion worked out by Nguyen [7] concerning to the quality of expert knowledge. However, when the number of collective members is small, the relationship between C2 values and the quality of collective knowledge is unclear. Meaning we can't give any statements about relationship between C2 values and the quality of collective knowledge. This problem will be mentioned in figure 3.For C3, C4 values, we can see that in all iterations the values of function C3 oscillate around the value 0.5 while the values of function C4 oscillates around 0.51. It is worth noting that these two functions are dependent on each other, as we mentioned in section 2.2. In this case, we still can't give any statements about relationship between C3 and Q as well as between C4 and Q.

In Figure 2, we show the results of collectives with smaller members (100) and the length of vector is 10. Again, we can see that the function C2 has the greatest impact on the quality of collective knowledge, although the relation seems to be less accurate than the case with 500 members. However, we can see more significant influence of functions C3 and C4 on the quality of collective knowledge. Slight increase of values of both functions C3 and C4 is reflected in the increase of the quality.

On the other hand, when the size of collective is small (Figure 3), we note that it is harder to determine the explicit relationship between the consistency C2 and the quality of collective knowledge than in the previous cases. Inversely, functions C3 and C4 show significant influence on the quality of collective knowledge. We can also clearly see that the smaller the size of collective, the greater the difference between values of functions C3 and C4.

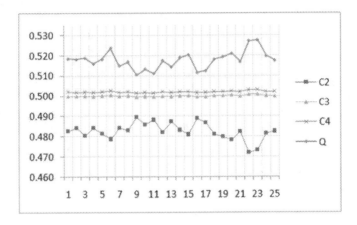

Fig. 1. 500 members of length 10

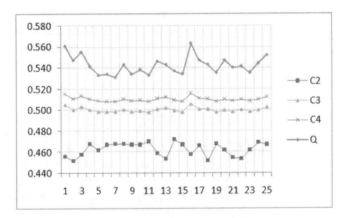

Fig. 2. 100 members of length 10

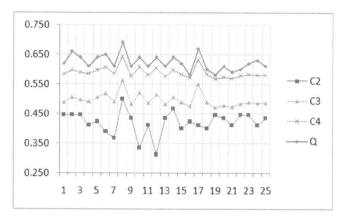

Fig. 3. 10 members of length 10

4.3   Statistical Analysis

In the previous section, we have some visual statements relationships between consistency degree and quality of collective knowledge through charting analysis. However, we couldn't evaluate how strong the relationships are? For this aim, statistical test is used to examine the relationship between the consistency function values and the quality of collective knowledge. We know that, the correlation coefficient reflects the strength and direction of the linear relationship between the two variables. The data from our experiments do not come from a normal distribution *(according to the Shapiro-Wilk tests)*. Therefore, we choose the Spearman correlation as the method to measure the relationship between consistency function values and the quality of collective knowledge.

As we can see from Table 3, almost experiments with collectives of size greater than or equal to 100 members there is a strong decrease in monotonic relationship between C2 and Q. This is similar what we observed from the previous section. That is when the C2 consistency increases, the quality of collective knowledge decreases. In addition, in the next column, the p-values are much less than 0.05. They are statistically significant and the relationship is reliable. Inversely, when the size of collective is relatively small to the length of binary vector, the relationship is different. It is contrary to previous statement (10 members of length 5) or weaker (100 members of length 15). According to the *p-values* we can state that the correlation coefficient in this case is not statistically significant. Therefore, in this situation we can't give any statement about the relationship between C2 and Q. For relationship between C3 and Q, in all situations the correlation coefficients are approximate or over 90%. It means that the relationship between C3 and Q is very strong. It also means the higher the C3 consistency, the better the quality of collective knowledge. Additionally, in the last column p-values in all situations are very small and much less than 0.05. Therefore, the correlation coefficients in these situations are statistically significant. The functions C3 and C4 are dependent on each other; therefore the relationship between C4 and Q is similar to the relationship between C3 and Q.

Table 3. Correlation coefficients of C2, C3 and Q

Length	Size	(C2, Q)	p-value	(C3,Q)	p-value
5	10	0.20918	0.3156	0.88093	6.216e-09
5	50	-0.94532	1.114e-12	0.93358	9.842e-12
5	100	-0.97761	< 2.2e-16	0.96645	4.497e-15
10	100	-0.46562	0.01899	0.89719	1.248e-09
10	500	-0.91375	1.8e-10	0.92743	2.644e-11
10	1000	-0.99441	< 2.2e-16	0.95476	1.319e-13
15	100	-0.21644	0.2987	0.92841	2.273e-11
15	500	-0.81722	6.182e-07	0.93823	4.37e-12
15	1000	-0.91278	2.035e-10	0.88919	2.836e-09

For more concrete, we consider relationships between consistency function values and the quality of collective knowledge as follows:

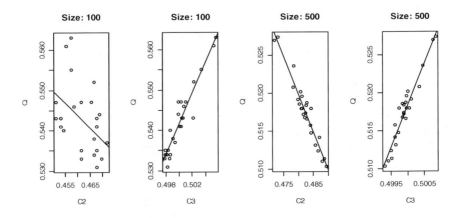

Fig. 4. Linear relationships between C2, C3 and Q

Figure 4 presents the linear relationship between consistency function values and the quality of collective knowledge with different size of collectives (100, 500) and the length of binary vectors is identical (10). It gives us clearer understanding about the relationship between the consistency function values and the quality of collective knowledge. Concretely, with function C2, the lower consistency function value, the better the quality of collective knowledge. Besides, when the number of collective knowledge is large enough, then the relationship between consistency function value and the quality of collective knowledge is stronger. Inversely, with C3 and C4, we can state that the higher the consistency value, the better the quality of collective knowledge. They are independent on the size of collective.

5   Conclusions and Future Works

In this paper, we have investigated analyzing influence of consistency degree on quality of collective knowledge using binary vector structure. Through different test cases, the experiments have shown that consistency degree have influence on the quality of collective knowledge. That is, the larger the average distances of collective, the better the quality of collective knowledge. It also means the hypothesis in Introduction is true. Besides, through experiments analysis we can see that when the number of members in a collective is large enough then the more different the opinions of collective members, the better is the credibility of collective knowledge.

The future works should be proving these relationships by using mathematical model, analyzing influence of consistency degree on the quality of collective knowledge with other knowledge structures, determining the number of collective members in which satisfies the statement the higher inconsistency, the more proper the collective knowledge. To the best of our knowledge, paraconsistent logics can be useful [5]. We also plan to use ontology structure to analyze these aspects [3].

References

1. Arrow, K.J.: Social Choice and Individual Values. Wiley, New York (1963)
2. Day, W.H.E.: The consensus methods as tools for data analysis. In: Bock, H.H. (ed.) Classification and Related Methods of Data Analysis, Proceedings of IFCS 1987, pp. 317–324. North-Holland (1987)
3. Duong, T.H., Nguyen, N.T., Jo, G.S.: A Method for Integration of WordNet-Based Ontologies Using Distance Measures. In: Lovrek, I., Howlett, R.J., Jain, L.C. (eds.) KES 2008, Part I. LNCS (LNAI), vol. 5177, pp. 210–219. Springer, Heidelberg (2008)
4. Li, T.: A general model for clustering binary data. In: Proceedings of the Eleventh ACM SIGKDD International Conference on Knowledge Discovery in Data Mining, Chicago, Illinois, USA, pp. 188–197. ACM (2005)
5. Nakamatsu, K., Abe, J.M.: The paraconsistent process order control method. Vietnam Journal of Computer Science 1(1), 29–37 (2014)
6. Nguyen, N.T.: Metody wyboru consensusu i ich zastosowanie w rozwiązywaniu konfliktów w systemach rozproszonych. Wroclaw University of Technology Press (2002)
7. Nguyen, N.T.: Advanced Methods for Inconsistent Knowledge Management. Springer, London (2008)
8. Nguyen, N.T.: Inconsistency of knowledge and collective intelligence. Cybernetics and Systems: An International Journal 39(6), 542–562 (2008)
9. Nguyen, N.T.: Processing inconsistency of knowledge in determining knowledge of collective. Cybernetics and Systems: An International Journal 40(8), 670–688 (2009)
10. Padula, M., Reggiori, A., Capetti, G.: Managing Collective Knowledge in the Web 3.0. Evolving Internet. In: First International Conference on INTERNET 2009 (2009)
11. Ridder, J.: Epistemic dependence and collective scientific knowledge. Synthese, 1–17 (2013)
12. Rolin, K.: Science as collective knowledge. Cognitive Systems Research 9(1-2), 115–124 (2008)

Evaluation of Domain Adaptation Approaches to Improve the Translation Quality

Ezgi Yıldırım[1] and Ahmet Cüneyd Tantuğ[2]

[1] Dep. of Computer Eng., Istanbul Technical University, Istanbul, Turkey
yildirimez@itu.edu.tr
[2] Dep. of Computer Eng., Istanbul Technical University, Istanbul, Turkey
tantug@itu.edu.tr

Abstract. This paper focuses on the usage of different domain adaptation methods to build a general purpose translation system for the languages with limited parallel training data. Several domain adaptation approaches are evaluated on four different domains in the English-Turkish SMT task. Our comparative experiments show that the language model adaptation gives the best performance and increases the translation success with a relative **9.25%** improvement yielding **29.89** BLEU points on multi-domain test data.

1 Introduction

Statistical machine translation is the most promising machine translation (MT) method with the availability of large scale parallel corpora. In general, the performance of a statistical machine translation (SMT) system is directly related to the amount of the training data in the parallel corpus. The amount of parallel data increases over the past years, however, the majority of the efforts are still concentrated on a limited number of languages such as English, German, Arabic, Chinese and etc. For the languages where no more parallel data can be found, the translation quality can be improved by other means which may be adding linguistic information to the translation process, employing better sentence, word alignment methods and utilizing domain adapted systems.

The basic intuition behind domain adaptation in MT is that a general domain translation system can translate sentences from various domains with an average quality; whereas, a translation system trained on a particular domain is able to perform better as long as the input sentences are in that domain. None the less, there is no standard way of domain adaptation implementation. Domain adaptation approaches in SMT suggest either combination of separate domain specific systems or adapting only translation model (TM) component or adapting only language model (LM) component. In this study, we focus on the effect of different domain adaptation techniques on English-Turkish SMT, where the size of the training data is relatively small.

© Springer International Publishing Switzerland 2015
D. Camacho et al. (eds.), *New Trends in Computational Collective Intelligence,*
Studies in Computational Intelligence 572, DOI: 10.1007/978-3-319-10774-5_2

SMT to or from Turkish[1] is a challenging task because of the agglutinative nature of Turkish. This very productive morphology causes excessively large number of surface forms which causes serious data sparseness problems. In order to build a sufficient translation model to represent the translation process, the training of a SMT system based on Turkish word forms requires much more parallel sentences than the amount required for other languages with simple morphologies. Contrarily, the amount of available parallel corpora in Turkish is very limited when compared to other language pairs. Domain adaptation may be effective in improving translation performance for the languages that are negatively influenced by the lack of sufficient training data.

The rest of the paper is structured as follows: In Section 2, we present a brief introduction of different domain adaptation techniques in MT. Section 3 describes a baseline system and the domain adaptation methods used in this study. The data set and the results of our experiments are given in Section 4. Finally, Section 5 involves some conclusions and possible future works.

2   Related Work

If there is need for average quality general translation systems, combining all data can be the solution to be applied first. However, domain adaptation is essential for high quality of general translation systems. A phrase-based statistical machine translation system comprises two main components: translation model (TM) and language model (LM). These models are also tuned on development data to optimize translation quality. Most of the domain adaptation works are performed on adaptation of these components to different domains. Koehn and Schroeder [8] compared domain adaptation methods applied on these components. They obtained best performance by using two translation models through the alternative decoding paths. They also made the inference that language model is a useful component for domain adaptation.

Sennrich [13] investigated adaptation methods of translation models, and proposed perplexity minimization of translation model as a method to set model weights in mixture modeling for domain adaptation task.

Banerjee et al. [1] received the support of classifiers to combine different domain-specific translation systems. For their case, combining domain-specific models with a domain classifier gave better translation success than alternative decoding paths used in a previous study [8].

In the literature, the factored model framework of Moses is also used for domain adaptation [10]. Domain information is represented as an additional factor, and using domain factors are achieved improvement on translation success of both in-domain and out-of-domain data.

Since bilingual data are expensive to obtain, monolingual resources are sent in for domain adaptation. In the research of Nakov [9], the phrase tables and lexicalized reordering models are enriched by using in-domain bi-texts augmented

---

[1] Turkish is spoken by approximately 160M native speakers according to UN statistics in 1990.

with monolingual sentence-level syntactic paraphrases on source language side. Bertoldi and Federico [2] generated a synthetic bilingual corpus by translating monolingual adaptation data with a phrase-based SMT system. They showed the contributions of these additional data on the different adaptation approaches.

3 Domain Adaptation Methods

In this section, the domain adaptation methods investigated in this research are introduced. For a fair comparison of domain adaptation methods, a baseline translation system is required. The most intuitive way of building a simple SMT system is using all the available parallel data. So, we have built a baseline SMT system on a parallel corpus that contains all available English-Turkish parallel sentences from all domains. This system shown in Figure 1 is referred as Combined Data System (CDS) since it is trained and tuned on the combined set of parallel sentences. The diversity in the training data makes the baseline system appropriate for general purpose translations with acceptable level of translation quality.

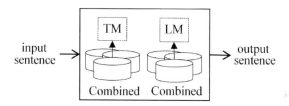

Fig. 1. Baseline System (Combined Data System)

3.1 Combining Domain-Specific Systems

Probably, the simplest way of constructing a domain adapted SMT system is training domain-specific SMT systems separately, and then combining those using a text classifier. The model proposed for this method is shown in Figure 2. Translation and language models of Domain Specific SMT (DS-SMT) system are trained and tuned on domain specific data. At run time, a classifier determines the domain of the input sentence[2] in the source language and chooses the appropriate DS-SMT system trained on that domain.

---

[2] This classification may also be on paragraph or document level instead of sentence level.

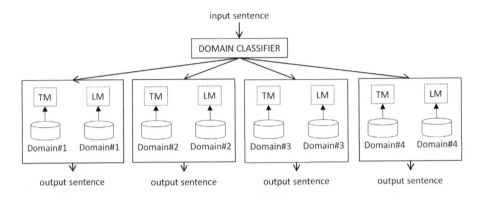

Fig. 2. Combining Domain-Specific Systems

Back-off Generalization. As the improved version of the previous method that relies solely on DS-SMT systems, we applied a modified model which takes advantages of multiple alternative decoding paths [3]. The motivation for using multiple alternative decoding paths is using a general translation model for the cases where no translation options can be found in the selected DS-SMT system. We combined two translation models (DS-SMT Translation Model and CDS Translation Model) using multiple decoding paths and back-off models for each domain. The resulting combined translation system seeks translation options (1) in the in-domain translation model (DS-SMT), and then (2) in the general translation model (CDS) if it could not find any possible translation options. The second translation table is the back-off table for unknown words and phrases in the first one. We also used two language models: an in-domain LM and general domain LM whose weights are set equal.

Multiple translation models are theoretically more proper than other single translation models, because the translation system has a chance to learn additional translation options if one model never gives an option for an input phrase. The same advantage can be gained by combining in-domain and general domain translation models together. Since LMs are more effective than TMs on domain adaptation task [6], we combined our baseline system with previously trained domain-specific systems instead of only TM-adapted systems in order to add general translation capabilities to DS-SMT systems with back-off method.

3.2   Using Domain as a Factor

Factored SMT gives the opportunity to involve additional information in the translation process. The main motivation of factors in SMT is increasing translation quality by incorporating additional linguistic properties, such as morphological, syntactic or semantic information. Likewise, domain information can also help translation process since the context usually determines word choices, word orders and etc. So, we have used domain information as a factor in the factored translation framework of Moses [7] to train a translation model involves domain type information.

As an example, lets consider the *interest*[3] word in English. In the domain of financial news, *faiz* is more probable than *ilgi* as a translation of *interest*. On the other hand, *ilgi* is probably a better translation option in the subtitles domain. Therefore, *"The credit interest rate is announced."* sentence should have a tendency for belonging to news domain because of "interest" word along with "rate" and "credit" words. These examples are shown in Table 1.

Table 1. Examples of factored translation options in different domains

```
P(interest|faiz, News)> P(interest|ilgi, News)
English: The credit interest rate is announced.
Turkish: Kredi faiz oranları açıklandı.

P(interest|faiz, Subtitles)< P(interest|ilgi, Subtitles)
English: The child didn't show an interest in his new toy.
Turkish: Çocuk yeni oyuncağına ilgi göstermedi.
```

As in the examples, the domain identifier is used in the translation system in conjunction with the surface forms. Since our focus is on domain adaptation, only surface word forms and domain information is used as factors. These translation factors are represented in Figure 3.

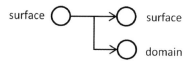

surface   surface

domain

Fig. 3. Translation factors of the system with domain factors

Translation factors are surface factors in both languages and domain identifier factor in the target language. There exists only one translation step without any generation steps. Obviously, translation performance can be improved by adding more linguistic information like stems, POS tags and etc., but that is beyond our scope in this work.

This model is subjected to test by multi-domain test data with surface form like in CDS, because the training is performed on only the surface form of source language.

---

[3] Interest : faiz (income return), ilgi (care, concern)

3.3   Combining Language Model Adapted Systems

A major drawback of DS-SMT systems are data sparseness because of their comparatively small in-domain training data size. Especially, agglutinate languages like Turkish seriously suffer from data sparseness problems due to the very large vocabulary size of surface forms generated by productive morphological structures. Therefore, splitting up the total training parallel data into domain-specific training sets may deteriorate the performance of DS-SMT system that does not cover sufficient translation options for input words or phrases. As stated by Daumé and Jagarlamudi [5], unseen words pose a major challenge for adapting translation systems to specific domains. On the other hand, language models can also be used as a domain adaptation component in a translation system. A large general domain translation model with the most available parallel data can be adapted to a specific domain by using an in-domain language model. This decreases the possibility to come across unseen words within a phrase.

On the contrary, progressively enlarging translation models can lead to overfitting problems for out-of-domain translation options of homonyms. However, in-domain language model penalizes improper options for its domain even if they are the most probable candidates in translation model. Since the translation system selects the most appropriate candidates for the specific domain, it will be able to produce jargon words and phrases in the domain.

For instance, *faiz* as a translation of *interest* is more probable than *ilgi* in news domain. By using domain specific language models, translation systems are inclined to translate phrases into that specific domain, therefore *faiz* is preferred in news domain.

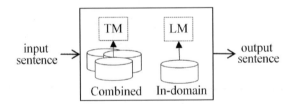

Fig. 4. Language Model Adapted System

As seen in Figure 4, the translation system is adapted to domain by using a general translation model with domain specific language models. Just like the model described in the previous section, a classifier is employed to select the appropriate domain for the input sentence.

4   Experiments

For our experiments, we used the Moses toolkit [7], an open source statistical machine translation system. We also used factored model framework of the system in some experiments (see Section 3.2 and Section 3.3). The SRILM language modeling toolkit [14] was used with Good-Turing discounting and interpolation. GIZA++ [11] is used for "grow-diag-final-and" symmetrized word-to-word alignment. We report our results with BLEU [12] translation evaluation metric. All experiments are carried on lowercased and tokenized data sets.

We evaluated our experiments first on oracle classifier test results, because we do not want to be affected by the success rate of the classifier. Oracle classifier score indicates the maximum score we would get if we know correct class of each input sentences. The clearest and reliable way of evaluation was done by using oracle classifier score, and at the end of experiments, the most successful method was reevaluated with real classifier described in Section 4.2.

4.1   Data

We made our research on four different domain corpora: news, literature, subtitles and web. The data in news [16] and subtitles domains are publicly available for all researchers[4]. The literature corpus is composed of novels, stories, politics and etc. [15] while the web corpus includes translations of web pages [17]. The number of sentences for each domain as well unique word counts for each data set is presented in Table 2 after eliminating sentences that has three or less words. For each domain, we built separate test and development sets of 2.5K sentences[5].

Table 2. Corpus Details

Domain	# of Sentences			Word Counts		Unique Word Counts	
	Training	Dev	Test	EN	TR	EN	TR
Literature	624,446	2.5K	2.5K	11,854,879	8,464,621	73,933	136,770
News	201,090	2.5K	2.5K	4,327,374	3,764,320	57,350	97,456
Subtitles	742,495	2.5K	2.5K	6,514,838	4,704,216	106,835	207,935
Web	141,467	2.5K	2.5K	3,083,162	2,591,270	72,367	106,369

4.2   Classifier Performance

For the experiments which require input sentence classification, we trained a linear text classifier. A multi-class Support Vector Machine (SVM) based classifier proposed by Crammer and Singer [4] is trained by using term frequency-inverse document frequencies (tf-idf) as features. An open source SVM toolkit for text categorization and analysis [18] is used to built an SVM classifier.

A number of tests to find the right features and pre-processing steps are employed. Eventually, bigram features with no stopword removal and no stemming give the best accuracy results.

---

[4] http://opus.lingfil.uu.se/, Last Accessed : June 2014
[5] These test sets are publicly available from
   http://ddi.itu.edu.tr/resources/domainData_en-tr.zip/

The training and development data from all domains (approximately 1.7M sentences) are used to train the classifier. To evaluate classifier performance, we run the classifier on sentences from four domain-specific test corpora. The classification results are reported in Table 3.

Table 3. SVM classifier accuracy for domain-specific and multi-domain test sets

Test Set	# of Test Inst.	Correct Predict	Wrong Predict	Accuracy(%)
Literature	2.5K	2405	95	96.2
News	2.5K	2168	332	86.72
Subtitles	2.5K	2345	155	93.8
Web	2.5K	2376	124	95.04
Multi-domain	10K	9294	706	92.94

As a classifier also introduces classification errors, we have employed our initial tests by an oracle classifier which always picks the right domain. After determining the most promising model, we re-run our test on this model only with the SVM classifier.

4.3   Domain Adaptation Results

Our first experiment utilizes a simple translation and a simple language model trained on the combination of all domain-specific data, which was called CDS previously. The result of this system is **27.36** in BLEU, and it is a baseline score for our domain adaptation models.

Prior to evaluating combination of DS-SMT systems, we have evaluated each DS-SMT system individually both with in-domain and out-of-domain test data. The results of these experiments are given in Table 4. For instance, the literature domain DS-SMT system achieved a **36.87** BLEU score with **2.5K** in-domain (literature) input test data, whereas its performance dramatically drops to **7.78** BLEU score on a **10K** general test data.

Table 4. Performances of domain-specific systems combination

Domain	Test Set	BLEU	N-gram precisions			
			1-gram	2-gram	3-gram	4-gram
Literature	In-domain	36.87	53.9	41.4	32.6	25.4
	Multi-domain	7.78	18.2	8.3	5.7	4.2
News	In-domain	17.17	46.6	23.3	12.5	7.3
	Multi-domain	3.98	22.8	5.9	2.1	1.0
Subtitles	In-domain	7.63	27.1	10.3	4.8	2.6
	Multi-domain	7.96	26.7	10.4	5.1	2.8
Web	In-domain	33.17	47.0	35.8	29.3	24.5
	Multi-domain	10.66	21.3	11.1	8.2	6.6

As expected, Table 4 reveals the fact that domain specific systems are able to produce high quality outputs for in-domain input and suffer from serious quality decreases for the out-of-domain input. These results show the importance of domain adaptation, and severity of using unsuitable translation models for out-of-domain sentences.

The combination of DS-SMT systems with an oracle classifier gives **27.92** BLEU score. Please note that the oracle classifier always predicts the right domain, so we have no classification mistakes.

As an extended version of combined DS-SMT systems, we used multiple decoding paths and back-off models of the Moses factored model framework. By using multiple decoding paths, we combined each DS-SMT system with the general purpose system (CDS) as a back-off model. For unknown phrases not found in the primary table (the domain specific TM), we perform a back-off search in the secondary table (a more general TM) up to four-grams.

The translation performances of four DS-SMT systems with back-off generalization are shown in Table 5. As seen in the table, this generalization seems to be effective for literature and web domains but produces worse results for news and subtitles domains. The reason for the deterioration might be using four-grams for back-off searches. The performance of DS-SMT systems on in-domain test sets (see 4) shows that the data for those domains is not as sufficient as to cover their test sets by comparison with other two domains, literature and web. Since they have the lower probability to catch a four-gram in the test data, they are more dependent to the CDS. Longer phrases in the general model prevail against shorter phrases in the domain specific systems even if the found longer phrases are not in the scope of the domain.

Table 5. Performances of domain-specific systems combination with back-off generalization

Domain	Test Set	BLEU	N-gram precisions			
			1-gram	2-gram	3-gram	4-gram
Literature	In-domain	44.10	67.2	53.8	44.0	35.8
News	In-domain	15.00	48.8	24.2	13.3	7.9
Subtitles	In-domain	5.90	26.5	9.9	4.9	2.8
Web	In-domain	38.16	56.7	44.4	37.4	32.1

The four DS-SMT systems with back-off generalization are used in conjunction with the oracle classifier to translate our 10K multi-domain test set and the overall performance of the system is **29.36** (see Table 7).

Using factored models for domain information is another domain adaptation method in this study. Basically, it is the extended version of our baseline (CDS) with the domain labels as factors. The training, development and test data are exactly same with the baseline system except that the target side surface forms are augmented with the domain labels. The domain identifiers are not used on the source side by assuming that the appropriate translation options for each domain can be learned during the training phase [10]. The main advantage of this method is that it does not require any classifiers. The experiment with factored translation gets **26.17** BLEU score. Domain factors take effect on the translation of word sequences including domain specific terms (means specific words or phrases which does not exist in other domain data). When a domain specific term is detected, surrounding words or phrases are translated in the same domain. Otherwise, in contrast with the motivation of using domain as a factor, the compulsory collocation of surface and domain factors together in target side scatters the overall probabilities into conditional probabilities, like in "*faiz* as the translation of *interest* if *faiz* belongs to the *news* domain". Therefore, the system opts the most probable translation among all conditional options even if it is not the best solution for the domain of test input. This deteriorates the success of translation system using domain information as additional factors.

Our last set of experiments are on domain specific language models. In these experiments, one large TM is trained on the combined data from all domains, and a separate domain-specific language model is used for each domain. We built four LM adapted SMT systems and a classifier to determine the system to be used. In order to investigate the in-domain performances of these systems, we have tested these four systems on in-domain data and the results are depicted in Table 6. When we compare LM adapted SMT versus the DS-SMT, it can be seen that LM adapted systems perform better than DS-SMT systems. For example, the DS-SMT system of literature domain achieves **36.87** BLEU score whereas a LM adapted SMT system of literature domain achieves **40.74**. The only exception is the news domain. As it is seen on the Table 2, news domain has the less unique word count which leads to incapable LMs in general.

Table 6. Performances of language model adapted domain systems combination

Domain	Test Set	BLEU	N-gram precisions			
			1-gram	2-gram	3-gram	4-gram
Literature	In-domain	40.74	58.8	45.7	36.3	28.3
News	In-domain	16.68	47.6	23.8	12.8	7.5
Subtitles	In-domain	8.11	26.6	10.5	5.5	3.4
Web	In-domain	36.88	52.1	39.9	32.7	27.3

Table 7 summarizes all experimental results. As can be seen in the table, language model adaptation outperforms other domain adaptation methods. From the point of view of classifier usage, we opt for using the SVM based real classifier only with the best performing method, i.e. (6) combining LM adapted systems.

The overall evaluation of different domain adaptation experiments shows that LMs are the most effective components for domain adaptation in English-to-Turkish SMT systems. While our (1) baseline gets **27.36** BLEU points, combined LM-adapted systems can get **29.89** BLEU score with a real classifier. The combination of LM-adapted systems yields 2.53 BLEU score gain (9.25% relative improvement) on multi-domain test data.

Table 7. Overall evaluation of various domain adaptation models

Adaptation Method	Test Set	Classifier	BLEU	Relative Improvement
(1) Baseline	Multi-domain	N/A	27.36	N/A
(2) Combining Domain-Specific Systems	Multi-domain	Oracle	27.92	2.05%
(3) Combining Domain-Specific Systems + Backoff Generalization	Multi-domain	Oracle	29.36	7.31%
(4) Using domain as a Factor	Multi-domain	N/A	26.17	-4.35%
(5) Combining LM-adapted Systems	Multi-domain	Oracle	**30.16**	**10.23%**
(6) Combining LM-adapted Systems	Multi-domain	SVM	29.89	9.25%

5 Conclusions

The experiments described in this paper present comparisons of various domain adaptation methods for English-to-Turkish statistical machine translation systems. We achieved the best SMT performance by language model adaptation. The translation performance can be increased from **27.36** to **29.89** BLEU points, which means **9.25%** relative improvement over the baseline SMT system performance. As a future work, we intend to increase the classifier success by applying document level and confidence level classification. Also using a language model controller can be investigated.

References

1. Banerjee, P., Du, J., Li, B., Kumar Naskar, S., Way, A., Van Genabith, J.: Combining multi-domain statistical machine translation models using automatic classifiers. Association for Machine Translation in the Americas (2010)
2. Bertoldi, N., Federico, M.: Domain adaptation for statistical machine translation with monolingual resources. In: Proceedings of the Fourth Workshop on Statistical Machine Translation, StatMT 2009, pp. 182–189. Association for Computational Linguistics, Stroudsburg (2009),
   http://dl.acm.org/citation.cfm?id=1626431.1626468
3. Birch, A., Osborne, M., Koehn, P.: Ccg supertags in factored statistical machine translation. In: Proceedings of the Second Workshop on Statistical Machine Translation, pp. 9–16. Association for Computational Linguistics (2007)
4. Crammer, K., Singer, Y.: On the algorithmic implementation of multiclass kernel-based vector machines. J. Mach. Learn. Res. 2, 265–292 (2002),
   http://dl.acm.org/citation.cfm?id=944790.944813
5. Daumé, I.H., Jagarlamudi, J.: Domain adaptation for machine translation by mining unseen words. In: Proceedings of the 49th Annual Meeting of the Association for Computational Linguistics: Human Language Technologies: Short Papers, HLT 2011, vol. 2, pp. 407–412. Association for Computational Linguistics, Stroudsburg (2011), http://dl.acm.org/citation.cfm?id=2002736.2002819
6. Foster, G., Kuhn, R.: Mixture-model adaptation for smt. In: Proceedings of the Second Workshop on Statistical Machine Translation, StatMT 2007, pp. 128–135. Association for Computational Linguistics, Stroudsburg (2007),
   http://dl.acm.org/citation.cfm?id=1626355.1626372

7. Koehn, P., Hoang, H., Birch, A., Callison-Burch, C., Federico, M., Bertoldi, N., Cowan, B., Shen, W., Moran, C., Zens, R., et al.: Moses: Open source toolkit for statistical machine translation. In: Proceedings of the 45th Annual Meeting of the ACL on Interactive Poster and Demonstration Sessions, pp. 177–180. Association for Computational Linguistics (2007)
8. Koehn, P., Schroeder, J.: Experiments in domain adaptation for statistical machine translation. In: Proceedings of the Second Workshop on Statistical Machine Translation, pp. 224–227. Association for Computational Linguistics (2007)
9. Nakov, P.: Improving english-spanish statistical machine translation: Experiments in domain adaptation, sentence paraphrasing, tokenization, and recasing. In: Proceedings of the Third Workshop on Statistical Machine Translation, StatMT 2008, pp. 147–150. Association for Computational Linguistics, Stroudsburg (2008), http://dl.acm.org/citation.cfm?id=1626394.1626414
10. Niehues, J., Waibel, A.: Domain adaptation in statistical machine translation using factored translation models. In: Proceedings of EAMT (2010)
11. Och, F.J., Ney, H.: A systematic comparison of various statistical alignment models. Computational Linguistics 29(1), 19–51 (2003)
12. Papineni, K., Roukos, S., Ward, T., Zhu, W.J.: Bleu: a method for automatic evaluation of machine translation. In: Proceedings of the 40th Annual Meeting on Association for Computational Linguistics, pp. 311–318. Association for Computational Linguistics (2002)
13. Sennrich, R.: Perplexity minimization for translation model domain adaptation in statistical machine translation. In: Proceedings of the 13th Conference of the European Chapter of the Association for Computational Linguistics, pp. 539–549. Association for Computational Linguistics (2012)
14. Stolcke, A., Zheng, J., Wang, W., Abrash, V.: Srilm at sixteen: Update and outlook. In: Proceedings of IEEE Automatic Speech Recognition and Understanding Workshop, p. 5 (2011)
15. Taşçı, Ş., Güngör, A.M., Güngör, T.: Compiling a turkish-english bilingual corpus and developing an algorithm for sentence alignment. In: International Scientific Conference Computer Science (2006)
16. Tyers, F.M., Alperen, M.S.: South-east european times: A parallel corpus of balkan languages. In: Proceedings of the LREC Workshop on Exploitation of Multilingual Resources and Tools for Central and (South-) Eastern European Languages, pp. 49–53 (2010)
17. Yıldız, E., Tantuğ, A.C.: Evaluation of sentence alignment methods for english-turkish parallel texts. In: First Workshop on Language Resources and Technologies for Turkic Languages, p. 64 (2010)
18. Yu, H., Ho, C., Juan, Y., Lin, C.: Libshorttext: A library for short-text classification and analysis. Tech. rep. (2013), http://www.csie.ntu.edu.tw/~cjlin/papers/libshorttext.pdf

Intelligent Problem Solving about Functional Component of COKB Model and Application

Van Nhon Do and Diem Nguyen

Vietnam National University HoChiMinh City (VNU-HCM),
University of Information Technology, Vietnam
{nhondv,diemnt}@uit.edu.vn

Abstract. Knowledge representation models and automated reasoning algorithms are the most important problems in designing knowledge-based systems in artificial intelligence, especially in intelligent problem solver (IPS). One of effective models is the Computational Object Knowledge Base model (the COKB model), which can be used to represent the total knowledge and to design the knowledge base of practical intelligent systems. However, besides the well-known knowledge components like concepts, relations, rules, there is another popular form of the knowledge, but has not been deeply researched. It is the functional knowledge component consisting of functions, relations, facts and rules and automated reasoning algorithms on functions. Consequently, in this paper, we will introduce the five-component COKB model and present in detail the knowledge representation method and automated reasoning algorithms for knowledge component about functions. In addition, this method has been used to design and implement an effective application, which can solve problems in solid geometry and produces step-by-step solutions.

Keywords: Knowledge representation, automated reasoning, intelligent problem solvers, knowledge-based systems.

1 Introduction

In many fields of science, there is a variety of methods to represent knowledge usually built based on mathematical foundations. The following are some methods of knowledge representation such as predicate logic, frames, classes, semantic networks, neural networks, conceptual graphs, deductive rules [1,2,3]. Although, these methods are very useful in many applications, they also have several disadvantages that lead to difficulties in representing certain knowledge domains systematically and naturally.

One of effective models used in IPS is the COKB model [4], which can be used for knowledge modeling, designing and implementing knowledge bases. Computational networks model (Com-Net) and the Networks of computational objects model (CO-Net) [4] are used for modeling problems in knowledge domains. These models are

---

* This research is funded by Vietnam National University HoChiMinh City (VNU-HCM) under grant number C2014-26-02

© Springer International Publishing Switzerland 2015
D. Camacho et al. (eds.), *New Trends in Computational Collective Intelligence*,
Studies in Computational Intelligence 572, DOI: 10.1007/978-3-319-10774-5_3

very appropriate tools for designing an inference engine of IPS. They have been used in constructing some practical IPS in education such as the program for studying and solving problems in plane geometry, analytic geometry algebraic problems [11,12], the program for solving problems in electricity, in inorganic chemistry, etc. In spite of many successful applications in practical knowledge domains, the function component and operator component of the COKB model have not been fully studied.

There are many current programs solving geometry automatically; however, besides their many advantages, they also have some limits. Firstly, there are some well-known tools such as Maple, Mathematica [13,14]. Nevertheless, they are just suitable for computing only; they cannot give step-by-step solutions for explaining the process of solving geometry problems. Subsequently, Wolfram|Alpha and Mathway websites [14,15] can produce step-by-step proofs. Nonetheless, they just solve very simple problems in analytic geometry. The next very effective program is Java Geometry Expert (JGEX) [9] based on Wu's method, the full-angle method and the deductive database method, etc. Since the important success of the Wu's method [6] introduced for mechanical theorem proving in geometry, hundreds of difficult geometry problems have been solved. However, the proofs generated by these algebraic methods differ from the traditional proof methods because of involving computations of hundreds polynomials. Thus, the proofs of those methods are not natural and easy to comprehensible. Besides the traditional algebraic methods such as Wu's method, the deductive database method [8] and volume method [7] can produce readable proofs and each step of the proofs has a clear geometric meaning. Though the algorithms used in JGEX are also very efficient, the knowledge representation by predicates is unnatural and it cannot perform the full range of aspects of geometry diverse knowledge. For example, the method cannot represent knowledge related to expressions, equations, computational relations between geometric objects. Besides, the program cannot solve a full variety of geometry problems, such as calculating the value of the geometric objects. On the other hand, the COKB model shows the efficiency of representing knowledge and solving many problems in geometry.

There are wide knowledge and many problems related to functional component in real knowledge domains, such as knowledge of plane geometry, solid geometry, knowledge of alternating current in physics. For example, line of intersection between two planes is a line, so Line of Intersection can be a function; the common perpendicular of two skew lines returns a line can be modeled as a function. Though there are many studies [4,5] mentioned the functional component of COKB model, no research presents a completed solution for these kind problems. An illustration of this is COKB model [4] performed the functional component of knowledge but not performed and handled computational relations, reasoning between functions. One more example of this is in the rule set of the model represented in the thesis [5], rules were represented in the form of functional equations, there are not represent in other forms of deductive rule. Therefore, many problems still need to research on functional knowledge component.

Hence, the aim of this paper is to consider the five-component COKB model and propose in detail the representation method and automated reasoning algorithms for knowledge component about functions. Using knowledge base designed based on this method, we constructed a model for problems and designed reasoning algorithms for solving problems automatically. Besides that, the model was applied to design the knowledge-based system for solving problems in solid geometry [10] and providing reasonable geometric solutions.

2   COKB Model

Computational Objects

The structure of a computational object (Com-object) modeled in [4] has four components (Attrs, F, Facts, Rules) where Attrs is a set of attributes, F is a set of equations called computation relations, Facts is a set of properties or events of object, and Rules is a set of deductive rules on facts.

For example, knowledge about a quadrilateral pyramid consists of elements (Points, Quadrilateral, etc) together with formulas and some properties on them. This object can be modeled as a class of Com-objects whose sets are as follows:

- o Attrs = {S, A, B, C, D: Point; h, S_{ABCD}, V: float } is the set of all attributes of the quadrilateral pyramid. In which: h is the height from the top S of the pyramid to the base ABCD, S_{ABCD} is the area of the base ABCD, V is the volume of the pyramid.

- o F = {$V = \dfrac{1}{3} h S_{ABCD}$ },

- o Facts = {["Coplanar", A, B, C, D], Not["Belong", S, plane[ABCD]], ["Skew", SA, BC], ["Skew", SA, DC], ["Skew", SB, AD], ["Skew", SB, DC], Not["Belong", A, plane[SBC]], Not["Belong", A, plane[SDC]], Not["Belong", B, plane[SAD]], Not["Belong", C, plane[SAB]], ...},

- o Rules = {{S, A, B, C, D: identified} $\Rightarrow$ {QuadrilateralPyramid[S, A, B, C, D]: identified}; ...}.

Components of COKB Model

The functional knowledge component represented by the COKB model consists of five components as follows:

$$(C, H, R, Funcs, Rules)$$

The meanings of the components are as follows:

- o C is a set of concepts of computational objects. Each concept in C is a class of Com-objects.
- o H is a set of hierarchical relations on the concepts.
- o R is a set of relations on the concepts.
- o Funcs is a set of functions.
- o Rules is a set of rules.

Facts in COKB Model

The COKB model for knowledge component about functions does not meet all of 11 kinds. The kinds of facts are as follows:

- **Fact of kind 1:** information about object kind. For example, SABC is a triangle pyramid, SABCD is quadrilateral pyramid.
- **Fact of kind 2:** a determination of an object or an attribute of an object. For example, Segment[A,B], Segment[A,B].length
- **Fact of kind 3:** a determination of an object or an attribute of an object by a value or a constant expression. For example, Segment[A, B].length = const
- **Fact of kind 4:** equality on objects or attributes of objects. For example: O1.c = Segment[A,B]
- **Fact of kind 5:** a dependence of an object on other objects by a general equation. For instance, Segment/ Segment = Segment / Segment
- **Fact of kind 6:** a relation on objects or attributes of the objects. For instance, ["Belong", Point, Segment]
- **Fact of kind 8:** a determination of a function by a value or a constant expression. For instance, Distance(Line, Line) = const, Angle(Point, Point, Point) = const
- **Fact of kind 9:** equality between an object and a function. For instance, Point = Point_of_Intersection (Line, Line), Line = OrthogonalProjection(Line, Plane)
- **Fact of kind 10:** equality between a function and another function. For instance, Angle(Plane, Plane)= Angle(Line, Line), Distance(Point, Plane)= Distance(Point, Line)

3   Modeling Problems and Algorithms

Networks of Computational Objects

A Network of computational objects (CO-Net) consists of a set of Com-objects, $O = \{O_1, O_2, ..., O_n\}$ and a set of facts between the objects in set O, $F = \{f_1, f_2, ... , f_m\}$. A computational relation f between attributes of objects or between objects is called a relation between the objects. The CO-Net is denoted by **(O, F)**.   On the CO-Net (O, F), we consider problems that to determine (or compute, prove) attributes in set G. The problems is denoted by the symbol H → G, H is the hypothesis and G is the goal of the problem.

Example: Let S.ABC be a triangular pyramid, and I, K be arbitrary points of edges SA, SC respectively such that IK is not parallel to AC. Determine the point of intersection of IK and plane (ABC).

The problem can be considered on the Network of Com-objects (O, F) as follows:

O = {SABC: Triangle Pyramid; S, A, B, C, I, K: Point}
F = {f1, f2, f3} consists of the following relations:
    o f_1: ["Belong", I, Segment[SA]]   {I belongs to SA}
    o f_2: ["Belong", K, Segment[SC]]   {K belongs to SC}
    o f_3: Not ["Parallel", Segment[IK], Segment[AC]] {IK is not parallel to AC}
And goal G = {Determine: Point of Intersection of IK and Plane[ABC]}

Algorithm

The rules in the knowledge domain of Solid geometry are extremely complicated. Moreover, searching for deductive rules to produce new facts is a significant technique to find solutions of problems. Thus, finding suitable reasoning algorithms is a challenging work. As a result, breadth-first forward chaining plays an essential role in analyzing and producing the facts. Although the breadth-first search is extremely expensive, performing the deductive reasoning for geometry reasoning works well. However, in the worst cases of using the breadth-first search to solve some difficult geometry problems, their closures are too large to be reached within reasonable time and the computer space. Therefore, speeding up the searching rules by using heuristic rules is considered to find goals of problems. However, the toughest technique is designing a deductive method with heuristics rules.

The following is some techniques used for reasoning in the system:

- First of all, it is necessary to divide the set Rules into some specific sets as follows: Rules for generating new points and new functions, Rules for leading to negative relations, Rules for identifying objects, Rules for casting objects, "Natural" Rules, Rules for sample problems
- Heuristic rules about sample rules: After recording the elements in hypothesis part and goal part, the program will apply Rules for sample problems to find whether any sample rule can be used.
- Heuristic rules about generating new points and functions: Generating new points and functions may lead to endless geometric objects and prevent the reaching of goals as well as slowing down the searching speed. In order to avoid generate too many new points and functions; the following introduces strategies to control the adding of new points and functions. Firstly, the algorithm finds rules to reach new facts without using generating rules. If the goal is achieved, the program terminates. Otherwise, the program will try to construct a new points and new functions. The program will repeat this process until either the goal is reached or it cannot apply any generating rules. Secondly, we will check whether new facts are found by those new points and functions. If they are found, the program will keep the new points and functions; otherwise, they will be deleted.

With the supplementation of these heuristic rules, the functional knowledge model resolve almost various kinds of solid geometry problems. Especially, the heuristic rules will speed up the process of finding a satisfactory result in the functional knowledge model. Furthermore, the solutions given by the model are more natural and closer to human thinking and solving.

The following algorithm with heuristic technique used in the functional knowledge model is to find solutions for problems modeled by $(O, F) \to G$.

- Step 1: Record the elements in hypothesis part and goal part.
- Step 2: Check the goal G. If G is obtained then go to step 7.
- Step 3: Find the sample problems can be applied.
- Step 4: Using heuristic rules to select a rule for producing new facts or new objects.
- Step 5: If the selection in step 3 fails then Searching for any rule, which can be used to deduce new facts or new objects.

- Step 6: If there is a rule found in step 3, 4, 5 then record the information about the rule, new facts in Solution, and new situation (previous objects and facts together with new facts and new objects), and go back to step 2. Else {searching for a rule fails} give conclusion: Solution not found, and stop.
- Step 7: Reduce the solution found by excluding redundant rules and information in the solution.

4   Application

In this section, we will present an application of the COKB model for knowledge component about functions. The application applied the knowledge representation model and inference techniques of COKB model to design a knowledge base and solve problems in Solid geometry. This program was implemented in Maple and the user interface was designed in C# environment. The program with an intuitive interface was tested and it produced good results.

4.1   Designing the Knowledge Base

The knowledge base of Solid geometry represented by COKB model consists of five components: (C, H, R, Funcs, Rules). The components of the model are listed as follows:

Set C of Concepts
The set C consists of concepts of Com-objects. There is a variety of concepts in Solid geometry, including "Point", "Segment", "Line", "Angle", "Plane", "Triangle", "Equilateral Triangle", "Isosceles Triangle", "Right Triangle", "Isosceles Right Triangle", "Quadrilateral", "Trapezoid", "Right Trapezoid", "Parallelogram", "Rhombus", "Rectangle", "Square", "Triangular Pyramid", "Regular Triangular Pyramid", "Quadrilateral Pyramid" and "Square Pyramid".

Set H of Hierarchical Relations on the Concepts C

There are hierarchical relationships among the concepts in set C and they can be represented with using Hasse diagrams. For example, the following Hasse diagram demonstrates the hierarchy on the concepts of Quadrilaterals.

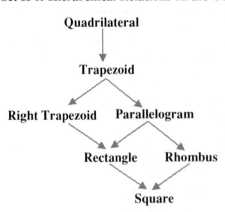

Fig. 1. The hierarchy on the concepts of Quadrilaterals

Set R of relations on Com-Objects
Set R contains various kinds of relations among Com-Objects. The following are some examples of relations: Relation Perpendicular between a Plane and a Plane, a Line and a Plane, a Line and a Line. Relation Distinct between a Point and Point, a Line and a

Line. Relation Belong between a Point and a Segment, a Point and a Line, a Line and a Plane. Relation Altitude between a Segment and a Triangle, a Segment and a Quadrilateral Pyramid.

Set Funcs of Functions on Com-Objects
The set Funcs consists of functions on Com-Objects, such as: Function *Line of Intersection* between two Planes returns a Line. Function *Point of Intersection* between two Lines, or between a Line and a Plane returns a Point. Function *Orthogonal Projection* of a Point onto a Line, or a Point onto a Plane returns a Point. Function *Orthogonal Projection* of a Line onto a Plane returns a Line. Function *Common Perpendicular* of two skew Lines returns a Line. Function *Distance* between a Point and a Line, a Point and a Plane, two Lines, a Line and a Plane, two Planes returns an integer number.

Set Rules of Rules
Rules is a set of deductive rules and each rule has the form "if <facts> then <facts>". There are some following examples:

- Rule 1: Three non-collinear points define a plane.
  $\{$A, B, C: Point; A, B, C are non-collinear $\Rightarrow$ Determine a plane $($ABC$)\}$

- Rule 2: If a point belongs to any lateral edge of a triangular pyramid then the point is not on the base of the triangular pyramid.
  $\{$SABC is a triangular pyramid; $I \in SA \Rightarrow I \notin Plane(ABC)\}$

- Rule 3: If a point is not in a plane then any line containing the point does not lie on the plane. $\{d : Line; A : Point; P : Plane; A \notin P; A \in d \Rightarrow d \not\subset P\}$

- Rule 4: Generated rule an orthogoral projection of a Point on a Line.
  $\left\{\begin{array}{l} d: Line, A: Point \Rightarrow Let\ I\ is\ orthogoral\ projection\ of\ A\ on\ d, and \\ AI \perp d, I \in d, AI \subset Plane(d, A) \end{array}\right\}$

- Rule 5: The property to recognize Isosceles Right Triangle: If a triangle has a right angle between the two equal edges then the triangle is the isosceles right triangle.
  $\left\{\triangle ABC; \widehat{BAC} = \dfrac{\pi}{2}; AB = AC \Rightarrow \triangle ABC\ is\ isosceles\ right\ triangle\ at\ A\right\}$

- Rule 6: The property to recognize an altitude of a Triangle.
  $\{\triangle ABC; AH \perp BC; H \in BC \Rightarrow AH\ is\ a\ altitude\ of\ \triangle ABC\}$

- Rule 7: The property of two perpendicular lines.
  $\left\{d, d1: Line; A, B, C: Point; d \perp d1; A = d \cap d1; B \in d; C \in d1 \Rightarrow \widehat{BAC} = \dfrac{\pi}{2}\right\}$

- Rule 8: If a point belongs to both two distinct lines, then the point is the point of intersection of both of them.
  $\{d, d1: Line; H: Point; H \in d; H \in d1 \Rightarrow H = d \cap d1\}$

- Rule 9: If a line perpendicular to a plane then that line will perpendicular to any line belonging that plane. $\{d, d1: Line, (P): Plane, d \perp (P), d1 \subset (P) \Rightarrow d \perp d1\}$

- Rule 10: If a line is perpendicular to each of two intersecting lines at their point of intersection, then the line is perpendicular to the plane determined by them

$$\left\{ \begin{array}{l} \text{d, d1, d2: Line; P: Plane} \\ d \perp d1, d \perp d2, d1 \in (P), d2 \in (P), d1 \parallel d2, d \not\subset (P) \Rightarrow d \perp (P) \end{array} \right\}$$

- Rule 11: The common perpendicular of two skew lines is the unique line intersecting both of them at right angles.

$$\left\{ \begin{array}{l} \text{d, d1, d2: Line; A, B: Point} \\ \text{d1 is skew to d2}, d \perp d1, d \perp d2, A = d \cap d1, B = d \cap d2 \\ \Rightarrow \text{d is common perpendicular of d1 and d2, and } Distance(d1, d2) = \text{AB} \end{array} \right\}$$

4.2   Modeling Problems

Almost problems in Solid geometry can be modeled by using *the Networks of Computational Objects*. The structure of the problem model is denoted by: $(O, F) \rightarrow G$

Example: Given a quadrilateral pyramid SABCD whose base is the square ABCD with side a, SA ⊥ plane(ABCD) and the length of the segment SA = a. Construct the common perpendicular and calculate the distance of two lines AD and SB.
This problem can be modeled as follows:

$$O = \left\{ \begin{array}{l} \text{[S,Point], [A,Point], [B,Point], [C,Point],} \\ \text{[D,Point], QuadrilateralPyramid[SABCD],} \\ \text{Square[ABCD]} \end{array} \right\} ,$$

$$F = \left\{ \begin{array}{l} \text{Segment[AB] = a,} \\ \text{Segment[SA] = a,} \\ \text{Line[SA]} \perp \text{plane[ABCD]} \end{array} \right\}$$

➔ Goal: $G = \left\{ \begin{array}{l} \text{Determine: CommonPerpendicular(Line[AD], Line[SB])}, \\ \text{Calculate: Distance(Line[AD], Line[SB])} \end{array} \right\}$

4.3   Illustrating for the Algorithm

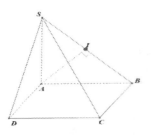

In this part, we will illustrate the algorithm presented in COKB model in solving problem given in the previous part "**Modeling Problems**".
Solution found by the program:
Note that, some steps of the following output solution is displayed in the human-readable form instead of using specification language. Therefore, the program helps users get easy to understand the solution.

1. $\text{plane}[ABCD] \Rightarrow AD \in \text{plane}[ABCD], AB \in \text{plane}[ABCD]$ **by rule: "The property of the concept Plane"**

2. $\text{Square}[ABCD] \Rightarrow AB \perp AD$ **by rule: "The property of the concept Square"**

3. $SA \perp \text{plane}[ABCD], AD \subset \text{plane}[ABCD] \Rightarrow SA \perp AD$ **by rule 9**

4. $D \not\subset \text{plane}[SBA] \Rightarrow AD \not\subset \text{plane}[SBA]$ **by rule 3**

5. $\text{plane}[SAB] \Rightarrow AB \in \text{plane}[SA\ B], SA \in \text{plane}[SAB]$ **by rule: "The property of the concept plane"**

6. $SA \perp AD, AB \perp AD, AB \subset \text{plane}[SAB], SA \subset \text{plane}[SAB], AD \not\subset \text{plane}[SAB]$
   $\Rightarrow AD \perp \text{plane}[SAB]$ **by rule 10**

7. $\text{Point}[A], \text{Line}[SB] \Rightarrow \text{Let } I = \text{OrthogonalProjection}(A, \text{Line}[SB]),$
   $AI \perp SB, I \in SB, AI \subset \text{plane}[SAB]$ **by rule 4**

8. $AD \perp \text{plane}[SAB], AI \subset \text{plane}[SAB] \Rightarrow AD \perp AI$ **by rule 9**

9. ["Distinct", $\text{Line}[AI], \text{Line}[SB]], I \in AI, I \in SB$
   $\Rightarrow I = \text{Point\_of\_Intersection}(\text{Line}[SB], \text{Line}[AI])$ **by rule 8**

10. ["Skew", $\text{Line}[SB], \text{Line}[AD]], AI \perp SB, AI \perp AD, A = \text{Point\_of\_Intersection}(\text{Line}[AD], \text{Line}[AI]), I = \text{Point\_of\_Intersection}(\text{Line}[SB], \text{Line}[AI])$
    $\Rightarrow \text{Line}[AI] = \text{CommonPerpendicular}(\text{Line}[SB], \text{Line}[AD]), \text{Distance}(\text{Line}[SB], \text{Line}[AD]) = \text{Segment}[AI]$ **by rule 11**

11. $SA \perp AB, A = \text{Point\_of\_Intersection}(\text{Line}[SA], \text{Line}[AB]) \Rightarrow \text{Angle}[SAB] = Pi/2$ **by rule 7**

12. $\text{Triangle}[SAB], \text{Angle}[SAB] = Pi/2, \text{Segment}[SA] = a, \text{Segment}[AB] = a$
    $\Rightarrow \text{IsoscelesRightTriangle}[ASB]$ **by rule 5**

13. $\text{IsocelesRightTriangle}[ASB], AI \perp SB, I \in SB$
    $\Rightarrow$ ["Altitude", $\text{Line}[AI], \text{IsoscelesRightTriangle}[ASB]]$ **by rule 6**

14. ["Altitude", $\text{Line}[AI], \text{IsoscelesRightTriangle}[ASB]]$
    $\Rightarrow \text{Segment}[AI] = \dfrac{\text{Segment}[SA] * \sqrt{2}}{2}$ **by rule in Isosceles Right Triangle Object**

15. $\text{Segment}[AI] = \dfrac{\text{Segment}[SA] * \sqrt{2}}{2}, \text{Segment}[SA] = a \Rightarrow \text{Segmen}[AI] = \dfrac{a\sqrt{2}}{2}$

16. $\text{Segmen}[AI] = \dfrac{a\sqrt{2}}{2}, AI = \text{Distance}(\text{Line}[SB], \text{Line}[AD])$
    $\Rightarrow \text{Distance}(\text{Line}[SB], \text{Line}[AD]) = \dfrac{a\sqrt{2}}{2}$

5   Conclusion and Future Work

This paper studied the five-component COKB model and proposed a knowledge representation method on functional knowledge component, consisting of functions, facts, relations and rules on functions. Subsequently, we modeled the problems and designed the automated reasoning algorithms to solve problems on this knowledge domain. It is a significant contribution to knowledge representation methods and knowledge processing techniques on knowledge component about functions.

Furthermore, we have implemented and designed an automatic solving program. The automated reasoning algorithm is very efficient and can solve successfully some kinds of problems in Solid geometry. The proofs produced by the program are generally readable and totally geometric. The program has been tested and received a good evaluation. Thus, this program is also a helpful educational software, and supports effectively for learning Solid geometry of students.

In the future, we will continue to complete COKB model to fully represent the knowledge component about functions. Besides, we will develop reasoning algorithms with heuristics rules and other inference algorithms on functional knowledge component for solving a wide range of problems on other knowledge domains.

References

1. Russell, S., Norvig, P.: Artificial Intelligence – A modern approach, 3rd edn. Prentice Hall, by Pearson Education, Inc. (2010)
2. van Harmelem, F., Vladimir, Bruce: Handbook of Knowledge Representation. Elsevier (2008) ISBN: 978-0-444-52211-5
3. Sowa, J.F.: Architectures for Intelligent Systems. IBM Systems Journal 41(3), 331–349 (2002)
4. Van Do, N.: Intelligent Problem Solvers in Education. In: Koleshko, V.M. (ed.) Design Method and Applications, Intelligent Systems. InTech (2012) ISBN: 978-953-51-0054-6
5. Do, N., Pham, T.L.: Knowledge representation and algorithms for automatic solving integral problems. In: Proceeding of IEEE 2011 6th International Conference on Computer Science & Education (ICCSE 2011), Singapore, pp. 730–735 (2011) ISBN: 978-1-4244-9716-4
6. Wu, W.-T.: On the Decision Problem and the Mechanization of Theorem in Elementary Geometry. Scientia Sinica 21, 159–172 (1978) Also in: Automated Theorem Proving: After 25 years, A.M.S., Contemporary Mathematics, 29, pp. 213–234 (1984)
7. Chou, S.-C., Gao, X.-S., Zhang, J.-Z.: Automated Production of Traditional Proofs in Solid Geometry. J. Autom. Reasoning 14(2), 257–291 (1995)
8. Chou, S.C., Gao, X.S., Zhang, J.Z.: A Deductive Database Approach To Automated Geometry Theorem Proving and Discovering. Journal of Automated Reasoning 25(3), 219–246 (2000)
9. Ye, Z., Chou, S.-C., Gao, X.-S.: An Introduction to Java Geometry Expert. In: Sturm, T., Zengler, C. (eds.) ADG 2008. LNCS, vol. 6301, pp. 189–195. Springer, Heidelberg (2011)
10. Doan, Q., Nhu, C.V., Khac, B.P., Ta, M.: Textbook 11th Grade Geometry. Publisher Vietnam Education (2010)
11. Nhon Van Do, A.: Program for Studying and Solving Problems in Plane Geometry. In: Proceedings of International on Artificial Intelligence 2000, Las Vegas, USA, pp. 1441–1447 (2000)

12. Do, N., Nguyen, H.: A Reasoning method on Knowledge Base of Computational Objects and Designing a System for automatically solving plane geometry problems. In: Proceeding of World Congress on Engineering and Computer Science (WCECS 2011), San Francisco, USA, pp. 294–299 (October 2011) ISBN: 978-988-18210-9-6

13. Bernardin, L., Chin, P., DeMarco, P., Geddes, K.O., Hare, D.E.G., Heal, K.M., Labahn, G., May, J.P., McCarron, J., Monagan, M.B., Ohashi, D., Vorkoette, S.M.: Maple Programming Guide. In: Copyright © Maplesoft, a division of Waterloo Maple Inc. (2014), `http://www.maplesoft.com`

14. Mathematica software, Wolframalpha website. In: Wolfram Media Inc., `http://www.wolfram.com/mathematica/`, `http://www.wolframalpha.com/`

15. Bagatrix Math Software Technology ©2010 Bagatrix, Inc. (2010), `http://www.mathway.com`

Reducing Model of COKB about Operators Knowledge and Solving Problems about Operators[*]

Van Nhon Do and Hien D. Nguyen

Vietnam National University HoChiMinh city (VNU-HCM),
University of Information Technology, Vietnam
{nhondv,hiennd}@uit.edu.vn

Abstract. Knowledge representation plays a very important role for designing knowledge base systems as well as intelligent systems. Nowadays, there are many effective methods for representing such as: semantic network, rule-base systems, computational network. Computational Objects Knowledge Base (COKB) can be used to represent the total knowledge and design the knowledge base of systems. In fact, a popular form of knowledge domain is knowledge about operations and computational relations, especially computational knowledge domain, such as: Linear Algebra, Analytic Geometry. However, COKB model and the other models have not solved yet some problems about operators: specification of operator, properties of operator, reducing an expression. In this paper, we will present a reducing model of COKB. This model, called Ops-model, represents knowledge about operators between objects and solve some problems related to these operators. Through that, the algorithms for designing inference engine of model have been built up. Moreover, Ops-model has been applied to specify a part of knowledge domain about Direct Current (DC) Electrical Circuits and construct a program for solving some problems on this knowledge domain.

Keywords: knowledge representation, knowledge-base systems, intelligent problem solver, automated reasoning.

1 Introduction

Knowledge representation is the core of knowledge engineering as well as intelligent systems. There are many methods for representing such as: semantic network, rule-base systems be presented in [1, 2], ontology in book [3]. In real application, a popular component of knowledge domain, especially computational knowledge domain, is knowledge component about operations. Operators between objects of knowledge domain are necessary concepts, they help for representation this knowledge exactly. Nowadays knowledge models have been proposed and applied in those knowledge domains although they have limitations:

---

[*] This research is funded by Vietnam National University HoChiMinh City (VNU-HCM) under grant number C2014-26-02.

D. Camacho et al. (eds.), *New Trends in Computational Collective Intelligence*,
Studies in Computational Intelligence 572, DOI: 10.1007/978-3-319-10774-5_4

Base on semantic network, in [4], author buit Multilayered Extended Semantic Networks (abbreviated MultiNet), which are one of the few knowledge representation paradigms along the line of Semantic Networks with a comprehensive, systematic and publicly available document. As with other semantic networks, nodes in Multinet represent concepts in knowledge, and arcs between these nodes represent relations between concepts. However, computational relations between concepts was not mentioned in MultiNet.

On the other research, authors built mathematical structure of knowledge representation based on extension rules, which studied the methods to solve contracdiction problems with formalized model, in [9]. Nevertheless, extension rules was not effective for representing real knowledge domain and this model do not include operators.

As the results in [5, 6], an algebra structure for concepts of knowledge has been constructed. This structure represents components of concepts and relations on them. It also express operators between concepts. However, inference rules is a very important component of knowledge base was not mentioned in this structure.

COKB model has been performed many kinds of practical knowledge and used to construct intelligent systems (see [7]). COKB model with reasoning method base on Sample Problem [8, 10] has been applied to construct some practical intelligent problem solvers such as the program for solving problems in plane geometry, analytic geometry. Nevertheless, operators components was not suitably studied in COKB. Some problems related to operators was not mentioned such as: sepcification of operator and problems on operators.

In this paper, base on COKB model, a reducing model of it has been presented. This model, called *Ops-model*, represents knowledge about operators. The foundation of model includes concepts, operators, and inference rules of knowledge. Ops-model refers to operators between objects and some characteristics of operator: commutative, associative, identity. Classes of problems about operator knowledge will be also researched in this model. Though that, the algorithms for designing inference engine of model have been built up. Moreover, Ops-model has been applied to specify a part of knowledge domain about Direct Current (DC) Electrical Circuits and construct a program for solving some problems on this knowledge domain. The solution given by this program is step-by-step, natural, and has reasoning like human.

2   Knowledge Model about Operators

2.1   Computational Objects Knowledge Base

Definition 2.1: The model for knowledge bases of computational objects (COKB model) consists of six components:

(C, H, R, Ops, Funcs, Rules)

The meanings of the components are as follows: **C** is a set of concepts of computational objects. Each concept in C is a class of Com-objects, definition about them was proposed in [7]. **H** is a set of hierarchy relation on the concepts, and it can be considered as the Hasse diagram for that relation. **R** is a set of relations on the concepts, and in case a relation r is a binary relation it may have properties such as reflexivity, symmetry, etc. The set **Ops** consists of operators on **C**, this component represents a part of knowledge about operations on the objects. The set **Funcs** consists of functions on Com-objects. **Rules** is the set of rules, they represent statements, theorems, principles, formulas, and so forth.

Operators component is very important in knowledge domains, especially computational knowledge domains. However, COKB model has not been yet address the problems related to operators such as: specification of operator, rules for determining operator and problems on operators. Base on COKB, model of operator knowledge will be constructed. This model is an reducing model of COKB, it will study and solve some problems related to operators.

2.2   Knowledge Model about Operators

Knowledge about operators between objects plays an important role in real knowledge domains, especially computational knowledge. In this section, a reducing model of COKB about operators has been presented. This model include concepts, operators, and inference rules of knowledge.

Definition 2.2: A *knowledge model about operators*, called Ops-model, consists of three components:

$$\mathcal{K} = (\textbf{C, Ops, Rules})$$

In which:

- **C** is a set of concepts. Each concept is a kind like computational object with behaviors for solving problems on object.
- **Ops** is a set of operators. Each operator is a binary mapping, we consider the properties of it are: commutative, associative, identity.
- **Rules** is a set of inference rules. Rules in this model is one of two forms: deductive rules and equation rules.

C- set of concepts
Each concept in C is a class of objects that is modeled by:

$$(\textit{Attrs, EqObj, RulesObj})$$

Attrs is a set of attributes, they are variables have real values.

                                        ("real values" mean real numbers)

EqObj is a set of equations between of attributes.

RulesObj is a set of deductive rules, each rule has the form: u $\longrightarrow$ v, which u $\subseteq$ Attrs, v $\subseteq$ Attrs and u $\cap$ v = $\emptyset$

Eg. 2.1: In knowledge domain about Direct Current (DC) Electrical Circuits, structure of CIRCUIT concept include:

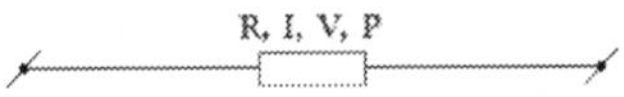

Fig. 1. A circuit in DC

$Attrs = \{R,V,I,P\}$: set of attributes of CIRCUIT
R,V,I,P: values of resistance, potential difference, current flow, power of circuit.

$$EqObj = \left\{ I = \frac{V}{R}, P = V.I, P = \frac{V^2}{R}, P = I^2R \right\}$$

$RulesObj= \{ \}$

A concept also has basic behaviors for solving problems on its attributes. Objects are equipped abilities to solve problems such as: 1/ Determines the closure of a set of object's facts. 2/ Executes deduction and gives answers for questions about problems of the form: determine some attributes from some other attributes. 3/ Executes computations. In this paper, we present the first problem about detemining the closure of a set of facts.

Definition 2.3: *The closure of a set of object's facts*
Let object Obj = (Attr, EqObj, RuleObj) is an obiect of concepts in C, and A is a set of facts related to attributes of Obj, these facts are classified as definition 2.5

a/ If e $\subset$ EqObj: e is a system of equations between k real variables $\{x_1,x_2,\ldots,x_k\} \subseteq$ Attr

e *can be applied* on A iff from facts of kind 4 and 5 in A, we have:
+ e can be solved to determine value of variables $\{x_1,x_2,\ldots,x_k\}$.

Let $e(A) = A \cup \{x_1,x_2,\ldots,x_k \}$

+ OR e produce a new relation between $\{x_1,x_2,\ldots,x_k\}$ in equation form.

Let $e(A) = A \cup subs(A, e)$ with subs(A,e) is a set of equations which substitutes some variables by values in A into system e.

b/ If r$\in$ RuleObj: r is a deductive rule has form: $u(r) \rightarrow v(r)$

r *can be applied* on A iff $u(r) \subseteq A$. Let $r(A) = A \cup v(r)$

c/ Let:
$$Obj.Closure(A) = \bigcup_{e \subset EqObj} e(A) \ \bigcup \ \bigcup_{r \in RuleObj} r(A)$$

Obj.Closure(A) is called the closure of set A by object Obj. Algorithm to determine Obj.Closure(A) will be presented as algorithm 3.1 in section 3.2

Ops-set of operators
This component represents operators between objects in knowledge domain. On the basis of operators, knowledge is equipped efficient computational methods for knowledge processing. An operator in this model is a binary mapping: $CixCi \rightarrow Ci$, with Ci is a concept in set C. Structure of an operator as followed:

operator-def ::= OPERATOR <name>
                ARGUMENT: argument-def+
                RETURN: return-def;
                PROPERTY: prob-type;
                ENDOPERATOR
argument-def ::= name, <name>: type

return-def ::= name : type
prob-type ::= commutative I associative I identity

Definition 2.4: Expression of objects in model is defined like this:
    expr ::= object I expr <operator> expr

Rules for determing an operator: Rules-set of Ops-model has been classified two kinds: kind for determining an operator and kind for inference rules of knowledge. The result of a operator is an object of concept in set C. The values of attributes of this object are determined based on rules for determining operator in Rules-set.
    Eg. 2.2: With CIRCUIT concept has been defined in Eg. 2.1, operator " + " performs the series connection beween two ciruits has been represented like this:
        OPERATOR < + > :
            ARGUMENT: CIRCUIT, CIRCUIT
            RETURN: CIRCUIT
            PROPERTY: commutative, associative
        ENDOPERATOR
Result of this operator has been determined base on rule R1 in section 4.2

Rules – set of rules
    In this paper, we consider an inference rule $r \in$ Rules is also one of the two forms:
        Rules = Rule$_{deduce} \cup$ Rule$_{equation}$
    • Form 1: $r \in$ Rule$_{deduce}$. In this form, r is an deductive rule has the form:
        $u(r) = \{f_1, f_2,...,f_p\} \longrightarrow \{q_1,q_2,...q_k\} = v(r)$
where f_i, q_i are facts, in which, kinds of fact has been classified as following:
Definition 2.5: Classify kinds of fact:

Kind of fact	Content	Specification	Example
1	information about object kind	<obj> : <type>	D : RESISTOR AB :CIRCUIT
2	a determination of an object or an attribute of an object	<obj> <obj>.<attribute>	
3	a determination of an object or an attribute of an object by a value or a constant expression	<obj> = <const expr> <obj>.<attribute> = <const expr>	D.V = 12 AB. I = 0.2
4	equality on objects or attributes of objects	<obj> = <obj> <obj>.<attribute> = <obj>.<attribute>	AB. I = D.I
5	equality on expressions of objects	<expression> = <expression>	AB = (D1 // D2) + D3

    • Form 2: $r \in$ Rule$_{equation}$. In this case, r is an equation of objects or attributes of objects, r has the form: $g(o_1, o_2,..., o_k) = h(x_1, x_2,..., x_p)$
where o_i, x_i are objects and g, h are expressions of objects.

3   Model of Problem and Algorithms

3.1   Model of Problem and Solution

Definition 3.1: Model of problemon Ops-model consists of three sets below.

$$O = \{O_1, O_2, \ldots, O_n\} \quad ; \quad F = \{f_1, f_2, \ldots, f_m\} \quad ; \quad G = \{g_1, g_2, \ldots, g_m\}$$

In the above model the set **O** consists of objects, **F** is the set of facts given on the objects, and **G** consists of goals. A goal of a problem may be the followings:
- Determine an object or some attributes of an object
- Compute a result of anexpression of objects.

The problem will be denoted by $(\mathbf{O}, \mathbf{F}) \rightarrow \mathbf{G}$

Definition 3.2: Let knowledge domain $\mathcal{K}$ = (C, Ops, Rules), object Obj = (Attr, EqObj, RuleObj) is an obiect of concepts in C, rule $r \in$ Rules and A is a set of facts. We have:

a/ Let Al_{Obj} = {f | f $\in$ A and f has attrributes of Obj}

$Obj(A) = Obj.Closure(Al_{Obj})$

$Obj(A)$ is called set of facts can be inferred from A by object Obj

b/ if r$\in$ $Rule_{deduce}$: r is a deductive rule has the form u(r) $\rightarrow$ v(r)

r *can be applied* on A iff u(r) $\subseteq$ A. Let r(A) = A $\cup$ v(r)

c/ if r$\in$ $Rule_{equation}$: r is an equation of k objects, r has form $g(x_1, x_2, \ldots, x_i) = h(x_{i+1}, x_{i+2}, \ldots, x_k)$ with x_i is an object.

r *can be applied* on A iff:

+ card(A $\cap$ $\{x_1, x_2, \ldots, x_k\}$) = k-1. Let r(A) = A $\cup$ $\{x_1, x_2, \ldots, x_k\}$

+ OR all objects in left(r) (or right(r)) were determined. Let r(A) = A $\cup$ r

r(A) is called set of facts can be inferred from A by rule r.

In which: card(X)   return the number of elements in set X

left(expr)   return the left hand side of expression expr

right(expr) return the right hand side of expression expr

Definition 3.3: Give a problem S = (O, F) $\rightarrow$ G on Ops-model. Suppose D = $[d_1, d_2, \ldots, d_m]$ is a list of elements which $d_j \in$ Rules or $d_j \in$ O. Denote: A_0 = F, A_1 = $d_1(A_0)$, $A_2 = d_2(A_1)$, ... $A_m = d_m(A_{m-1})$ và D(S) = A_m

A problem S is called *solvable* if there is a list D such that G $\subseteq$D(S). In this case, we say that D is a *solution* of the problem.

3.2   Algorithm for Solving Problem

Algorithm 3.1: Give an object Obj = (Attr, EqObj, RuleObj) as section 2.2 and A is a set of facts related to Obj. This algorithm is determine the closure of set A by Obj, Obj.Closure(A)

Result ← empty
NewEq ← EqObj
Solution_found ← false;
Do
 H ← Result
  Select f in EqObj and solve f:
    Update NewEq by f' as a new
    equation
    Update Result by f' as a new fact

Select{f, g} in EqObj to solve system e = {f,g}
    Update Result by varibles in f and g
Select deduce rule r in RuleObj and apply to
    produces new facts
    Update Result by v(r)
While (H != Result)
Result = Result ∪ NewEq
Obj.Closure(A) = A ∪ Result

Algorithm 3.2: Give the problem S = (O, F) → G as definition 3.1 on Ops-model, the solution of problem S has been found though these steps:

The following general algorithm represents one strategy for solving problems: forward chaining reasoning in which objects attend the reasoning process as active agents.

Solution ← empty list;
Solution_found ← false;
while not(Solution_found) **do**
  Use objects in set O and set of facts F to
    determine the closure of each objects
  if (object Obj produce the new facts) **then**
    Update the set of known facts, and add
      object Obj to Solution;
    if (goal G obtained) **then**
      Solution_found ← true;
    end if;
    continue;
  end if;
  Find a rule can be applied to produce new
    facts or new objects;
    (heuristics can be used for this action)

if (rule r found) **then**
  Use r to produce new facts or objects;
  Update the set of known facts and
    objects, and add r to Solution;
  if (goal G obtained) **then**
    Solution_found ← true;
  end if;
end if;
end do; { while }
if (Solution_found) **then**
  Solution is a solution of the problem;
else
  There is no solution found;
end if;

4   Application

In physical program of middle school in Vietnam, knowledge about Direct Current (DC) Electrical Circuitsis is very important. In this section, base on Ops-model, designing an intelligent software for solving problems on this knowlede domain has been presented. This program will solve problems automatically. Its solution is natural, step by step and simulate the way of human's solution.

4.1   Structure of Intelligent Problem Solvers

Structure of Intelligent Problem Solvers (IPS) system and processing to build them has been presented in [7, 10]. An IPS is also a knowledge base system, which supports searching, querying and solving problems based on knowledge bases; it has the structure of an expert system. We can design the system which consists of following components: knowledge base, inference engine, explanation component, working memory, knowledge manager, interface.

Knowledge base contain the knowledge for solving some problems in a specific knowledge domain. It must be stored in the computer-readable form so that the inference engine can use it in the procedure of automated deductive reasoning to solve problems stated in general forms. They can contain concepts and objects, relations, operators and functions, facts and rules.The Inference engine will use the knowledge stored in knowledge bases to solve problems, to search or to answer for the query. Moreover, it also have to produce solutions as human reading, thinking, and writing.

Based on structure of IPS system, knowledge base of system has been represented by model about operators, and Inference Engine has been designed by algorithms 3.1 and 3.2, we built an IPS in education which is called *the system for automatic solving problems in Direct Current Electrical Circuits.*

4.2   Design Knowledge Base

Base on knowledge about DC Electrical Circuits has been mentioned in [11], this knowledge domain can be represented by Ops-model as followed:

a) C–set of concepts

The set C consists of concepts such as "RESISTOR", "LAMB", "VariableResistor", "CIRCUIT", "CAPACITOR", "BATTERY".

Eg. 4.1: Structure of LAMB concept include:

$Attrs = \{$ ▪ R,V,I, P: values of resistance, potential difference, current flow, power of lamb

▪ V_{max} , I_{max} , P_{max}: maximum values of potential difference, current flow, power of lamb

▪ s: is a number performs three level of light, such as:

s = -1 : the light of lamb is low

s = 0 : the light of lamb is normal

s = 1 : the light of lamb is high $\}$

$$EqObj = \left\{ I = \frac{V}{R}, P = I^2 R, R = \frac{V^2_{max}}{P_{max}}, I_{max} = \frac{P_{max}}{V_{max}} \right\}$$

$RulesObj= \{$ if $(I < I_{max})$ or $(P < P_{max})$ or $(V < V_{max})$ then s = -1 ,

if $(I = I_{max})$ or $(P = P_{max})$ or $(V = V_{max})$ then s = 0 ,

if $(I > I_{max})$ or $(P > P_{max})$ or $(V > V_{max})$ then s = 1 $\}$

b) Ops–set of operators between concepts:

Operator	Meaning	Arguments	Return	Properties
+	Circuits can be connected by series connection	CIRCUIT x CIRCUIT	CIRCUIT	commutative associative
//	Circuits can be connected by parallel connection	CIRCUIT x CIRCUIT	CIRCUIT	commutative associative
o	Capacitors can be connected by series connection	CAPACICATOR x CAPACICATOR	CAPACI-CATOR	commutative associative
Ξ	Capacitors can be connected by series connection	CAPACICATOR x CAPACICATOR	CAPACI-CATOR	commutative associative

In this knowledge domain, RESITOR concept can also be seen as a circuit with one resistor, so operator + and // can be applied on RESITOR concept, these properties and rules of two operators are unchanged.

c) Rules–set of rules
Rules in this model are classified of two forms: deductive rules and equation rules
+ Some deductive rules in Rules-set:
R1: *Rule of series circuits*
  {D1, D2, AB: CIRCUIT, AB = D1 + D2}
  $\rightarrow$ {AB.V = D1.V + D2.V, AB.I = D1.I = D2.I, AB.R = D1.R + D2.R}

R2: *Rule of parallel circuits*
  {D1, D2, AB: CIRCUIT, AB = D1 // D2}
  $\rightarrow$ {AB.V = D1.V = D2.V,   AB.I = D1.I + D2.I, $\dfrac{1}{AB.R} = \dfrac{1}{D1.R} + \dfrac{1}{D2.R}$ }

R3: *Rule of series capacitors*
  {Cap1, Cap2, Cap: CAPACICATOR, Cap = Cap1 o Cap2}
  $\rightarrow$ {Cap.Q = Cap1.Q = Cap2.Q, Cap.V = Cap1.V + Cap2.V,
  $$\frac{1}{Cap.C} = \frac{1}{Cap1.C} + \frac{1}{Cap2.C}\}$$
R4: *Rule of parallel capacitors*
  {Cap1, Cap2, Cap: CAPACICATOR, Cap = Cap1 Ξ Cap2}
  $\rightarrow$ {Cap.Q = Cap1.Q + Cap2.Q, Cap.V = Cap1.V = Cap2.U,
  $$Cap.C = Cap1.C + Cap2.C\}$$
  + Some equation rules in Rules-set:
        R5: D1 + D2 // D3 = (D1 + D2) // D3
        R6: D1 // D2 + D3 = (D1 // D2) + D3
        R7: C1 o C2 o C3 = (C1 o C2) o C3
        R8: C1 Ξ C2 Ξ C3 = (C1 Ξ C2) Ξ C3

4.3   Design Inference Engine

Model of problem in this knowledge base about DC Electrical Circuits is defined as definition 3.1. Besides that, using algorithms 3.1 and 3.2, the inference engine has been built. This engine simulates the way of human thinking to find solution of practical problem.
  Eg. 4.2:
Problem S: Three resistors are connected   like   the   figure. Resistor D1 has value 30Ω, Resistor D3 has value 60Ω. The total current is 0.3 A, current though D3 is 0.2A .
What is the current though D1 ?

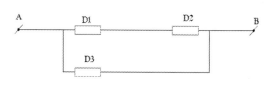

+ <u>Specification of Problem S:</u>

  O:= {D1, D2, D3: RESISTOR ; AB: CIRCUIT}

  F:= {AB = D1 + D2 // D3

    D1.R=30, D3.R=60, D3.I=0.2, AB.I =0.3}

  G:={D1.I}

+ <u>Solution of Program:</u>

1. D1 + D2 // D3 = (D1 + D2) // D3   by apply rule R5

2. Let [EF, CIRCUIT], EF = D1 + D2

  AB = D1 + D2 // D3

  D1 + D2 // D3 = (D1 + D2) // D3

    →AB = EF // D3

3. AB = EF // D3 → AB.I = EF.I + D3.I by apply "Rule of Paralle Circuit" R2

4. EF = D1 + D2 →EF.I = D1.I   by apply "Rule of Series Circuit" R1

5. From { D3.I = 0.2, AB.I = 0.3, AB.I = EF.I + D3.I} → EF.I = 0.1

            by "Solve equation"

6. From { EF.I = 0.1, EF.I = D1.I} → D1.I = 0.1

5   Conclusion and Future Work

COKB model is very useful and suitable for representing knowledge. We research an restricted form of COKB to built knowledge model about operators, *Ops-model*. On this model, some problems related to operators has been studied and solved, such as: specification of operator, determining result of operator and solving some kinds of problem on operators. Besides that, algorithm for finding solution of problems on this model has been designed. Ops-model has been applied for representing some knowledge domains. In this paper, we present using this model to represent a part of knowledge domain about DC Electrical Circuit and designing a program to solve problems on this knowledge domain. The solution given by this program is step-by-step natural, precise and has reasoning like human.

In the future, we will research to complete knowledge model about operators, some problems on operators have to be continued: reducing expressions, solving system of equations between objects. The improvement model can be applied to design knowledge base of DC Electrical Circuit is more completely and solve higher level problems. Moreover, we can apply this model to design knowledge base on other knowledge domains, such as: Physics (AC Electrical circuit, mechanics, optical) and Linear Algebra.

References

1. van Harmelem, F., Lifschitz, V., Porter, B.: Bruce: Handbook of Knowledge Representation. Elsevier (2008)
2. Russell, S., Norvig, P.: Artificial Intelligence – A modern approach, 3rd edn. Prentice Hall (2010)

3. Sowa, J.F.: Knowledge Representation: Logical, Philosophical and Computational Foundations. Brooks/Cole (2000)
4. Helbig, H.: Knowledge Representation and the Semantics of Natural Language. Springer, Berlin (2006)
5. Tian, Y., Wang, Y., Hu, K.: A Knowledge Representation Tool for Autonomous Machine Learning Based on Concept Algebra. In: Gavrilova, M.L., Tan, C.J.K., Wang, Y., Chan, K.C.C. (eds.) Transactions on Computational Science V. LNCS, vol. 5540, pp. 143–160. Springer, Heidelberg (2009)
6. Wang, Y.: On Concept Algebra and Knowledge Representation. In: Proceeding of 5th IEEE International Conference on Cognitive Informatics (ICCI) (2006)
7. Do, N.: Intelligent Problem Solvers in Education: Design Method and Applications. In: Koleshko, V.M. (ed.) Intelligent Systems. InTech (2012)
8. Do, N., Nguyen, H.: A Reasoning method on Knowledge Base of Computational Ojects and Designing a System for automatically solving plane geometry problems. In: Proceeding of World Congress on Engineering and Computer Science 2011 (WCECS 2011), San Francisco, USA, pp. 294–299 (October 2011) ISBN: 978-988-18210-9-6
9. Yang, C., Cai, W.: Knowledge Representations based on Extension Rules. In: Proceedings of the 7th World Congress on Intelligent Control and Automation, Chongqing, China (2008)
10. Do, N.V., Nguyen, H.D., Mai, T.T.: Designing an Intelligent Problems Solving System Based on Knowledge about Sample Problems. In: Selamat, A., Nguyen, N.T., Haron, H. (eds.) ACIIDS 2013, Part I. LNCS, vol. 7802, pp. 465–475. Springer, Heidelberg (2013)
11. Vietnam Ministry of Education and Training, Textbook and workbook of Physics. Publisher of Education (2012)

Context Identification of Scientific Papers via Agent-Based Model for Text Mining (ABM-TM)

Moamin A. Mahmoud[1], Mohd Sharifuddin Ahmad[1],
Mohd Zaliman M. Yusoff[1], and Aida Mustapha[2]

[1] College of Information Technology,
Universiti Tenaga Nasional, Kajang, Selangor, Malaysia
[2] Faculty of Computer Science & Information Technology
Universiti Putra Malaysia, Serdang, Selangor, Malaysia
{moamin,sharif,zaliman}@uniten.edu.my, aida@fsktm.upm.edu.my

Abstract. In this paper, we propose an agent-based text mining algorithm to extract potential context of papers published in the WWW. A user provides the agent with keywords and assigns a threshold value for each given keyword, the agent in turn attempts to find papers that match the keywords within a defined threshold. To achieve context recognition, the algorithm mines the keywords and identifies the potential context from analysing a paper's abstract. The mining process entails data cleaning, formatting, filtering, and identifying the candidate keywords. Subsequently, based on the strength of each keyword and the threshold value, the algorithm facilitates the identification of the paper's potential context.

Keywords: Text Mining, Data Mining, Agent-based Modelling, Context Identification.

1 Introduction

Hotho et al. [1] stated that text mining was initially mentioned by Feldman and Dagan [2], who discussed the machine-supported analysis of text, by exploiting several techniques of information retrieval, information extraction and natural language processing (NLP) [1, 2]. However, the term text mining is coined from the concept of data mining.

Data mining involves the sifting of data records in databases to identify significant patterns that are useful for a decision-making process [3]. Among the data mining tasks such as classification or clustering, association rule mining is one particular task that extracts desirable information structures like correlations, frequent patterns, associations or causal between sets of items in transaction databases or other data stores [3, 4]. Association rule mining is widely used in many different fields like telecommunication networks, marketing, risk management, inventory control and others [3]. The association rule mining algorithm discovers association rules from a given database such that the rule satisfies a predefined value of support and confidence. The aim of using support and confidence thresholds is to ignore those

© Springer International Publishing Switzerland 2015
D. Camacho et al. (eds.), *New Trends in Computational Collective Intelligence*,
Studies in Computational Intelligence 572, DOI: 10.1007/978-3-319-10774-5_5

rules that are not desirable, because the database is huge and users care about those frequently occurring patterns only [5].

The difference between data mining, association rule mining, and text mining is that in text mining, patterns are extracted from natural language text, but data mining and association rule mining patterns are extracted from databases [6]. According to Chiwara et al. [6], the text mining process steps are as follows:

- Text: This represents the given target document for mining which is in text format.
- Text processing: This step involves text clean up, format, tokenize and so on.
- Text transformation (attribute generation): Generating attributes from the given processed text.
- Attribute selection: Selecting attributes for mining because not all generated attributes are suitable for mining.
- Data Mining (Pattern discovery): Mining the selected attributes and extracting desirable patterns.
- Interpretation and evaluation: It is about what next, i.e. terminate, results well-suited for application at hand and so on.

Text mining is used by researchers to discover a particular desired information from unstructured and huge textual data [7]. Researchers applied their techniques in many domains such as to discover the useful knowledge for enterprise decision, scientific papers to handle information retrieval [7], medical data set to extract useful knowledge [8], news articles for Stock Price Predictions [9]. The literature in this area mainly focus on developing text mining techniques [7, 8, 9, 10, 11], but very few focus on developing a reasoning model that exploits text mining techniques to facilitate and accelerate extracting useful knowledge [12, 13]. However, some researchers exploit multi-agent systems as a reasoning model for this need [4, 5].

In this work, we propose a framework, which is called the Agent-based Model for Text Mining (ABM-TM) Framework, to extract keywords for identifying the potential context of a scientific paper. To do so, we develop an algorithm that mines textual data of scientific papers and identify the papers' potential contexts. In this framework, we exploit a multi-agent system to create a collaborative model that facilitates knowledge sharing. The motivation of this work stems from the problem of finding exact papers when students or researchers wish to find relevant papers from online search engines. Usually those engines do not give accurate results for the sought papers. Consequently, this paper aims to achieve three targets as follows:

- Facilitate Search: By developing a multi-agent system (MAS) model with an interface that handles interactions between an agent and its human counterpart. Such interface enables a user to find papers even if he/she is a beginner and does not have any basic skills about searching online because his/her agent assists him/her in the search.
- Accelerate Search: To achieve this target, we propose a Global Knowledge Base (GKB) to be as a store for all papers found by the multi-agent system (MAS). The GKB is structured based on the Case-based Reasoning (CBR) method.

Any new papers found from the WWW, the agent adds them to the GKB as a case structure. Those cases can be used by other agents if they have similar search parameters from other users. Over time, large number of cases are added to the GKB that potentially saves agent's time in searching and showing results, which eventually accelerate the search.

- Accurate Search: To get accurate results, we propose a Text Mining Algorithm (TMA) to identify the potential context. The TMA is inspired by the methods of data mining and association rule mining. It is developed from these methods with other processes to implement text mining. Figure 1 shows the proposed abstraction of the context identification process.

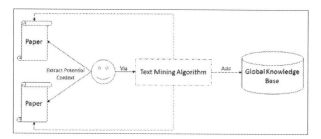

Fig. 1. Abstraction of Context Identification Process

This paper presents the work-in-progress of our work in search and context identification. It presents a framework and the associated processes of MAS, GKB, and TMA and leaves the implementation and results for future work. We present these processes as layers of the ABM-TMA framework.

The next section dwells upon the related work on text mining techniques. In Section 3, we introduce our framework. Section 4 presents the Text Mining Algorithm (TMA) and Section 5 concludes the paper.

2   Related Work

Reading documents to gain useful knowledge becomes a very time consuming task due to the fast and rapid growth of available information in texts. Therefore, it is useful and beneficial to have an automated knowledge discovery system for mining and extracting useful knowledge from texts [12].

Liu [8] explored text mining to extract multi-view heterogeneous data from an extensive publication database of Web of Science (WoS). In order to facilitate the scientific mapping that is useful for observing and identifying new trends in various scientific fields, they introduced a hybrid strategy named graph coupling. Their strategy identifies the clusters number and offers a top-down hierarchical analysis, which is suitable for practical applications.

Jusoh and Alfawareh [12], proposed an intelligent knowledge mining processor architecture, which is based on text mining and data mining approaches, and a multi-agent system. The architecture is designed using agent systems where each process in the architecture is presented as an agent. In the text mining approach, by exploiting data extraction and natural language processing techniques, they extract entity and concepts from the texts. In the data mining approaches, they apply classification and visualization techniques to classify the extracted concepts.

Zhong et al. [11] presented a pattern discovery technique. The technique includes pattern deploying and pattern evolving processes, to enhance the effectiveness of using and updating discovered patterns and observing relevant and useful information.

3   A Proposed Framework (ABM-TM)

In proposing the framework, we first define the following terms that relate to the framework.

Definition 1: A *keyword* is any significant word or phrase that is used to describe or represent the content of a document. A keyword or a number of keywords defines the context of a document.

Definition 2: A context is the meaning of a document's content as perceived by a reader of the document.

Definition 3: A keyword is attributed by its strength. A *keyword strength* is the degree to which the keyword expresses the meaning of a document's content. If a keyword strength of a document is greater than an assigned threshold value (given by a user), the document then meets the user's need.

Definition 4: A *threshold* is the minimum acceptable strength of a paper's keyword. For example, a user assigns a 70% threshold value for papers with the keyword 'agent' indicates that papers which return less than 70% of the keyword strength is not selected.

Definition 5: An *identity keyword* is the strongest keyword in a paper. The term identity is used because other keywords' strengths are measured based on this keyword.

Definition 6: A *potential keyword* is a paper's keyword which has a threshold value of less than or equal to the discovered keyword strength in a paper.

Definition 7: A *weak keyword* is a paper's keyword which has a threshold value of greater than the discovered keyword strength in a paper.

Definition 8: The *Global Knowledge Base* (GKB) is the store of all selected papers that have been identified by agents. The data of the GKB is structured based on the structure of the case-based reasoning (CBR) data.

The proposed framework introduces the concept of agent-based model for text mining. It attempts to find papers that meet users' needs through text mining. An agent gets a set of specific information from a user and by using text mining, it identifies the potential context of papers from the WWW and selects the relevant papers according to the user's input information. As shown in Figure 2, a user first input the desired keywords (1) and the threshold for each input keyword (2). An agent subsequently reads the inserted data (3) and starts the search within the Global Knowledge Base (KGB) (4). If there are matching cases, the agent shows the results to the human counterpart as initial results (5). For more results, the agent searches the World Wide Web (WWW) (6). It runs a text mining algorithm (7) to filter the discovered papers and assign the papers as cases for the user (8). Having discovered the papers, the agent adds them all as new cases to the GKB (9) and shows the cases as more results to the user (10).

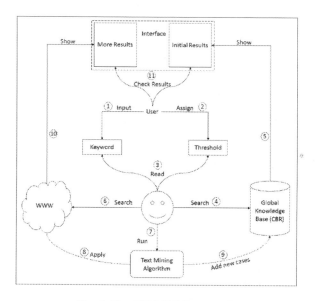

Fig. 2. The ABM-TM Framework

4   The ABM-TM Layers

The ABM-TM framework consists of three layers, which are interaction layer, knowledge sharing layer, and extraction layer (as shown in Figure 3).

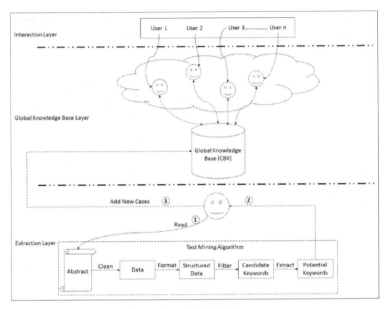

Fig. 3. The ABM-TM Layers

4.1   The Interaction Layer

The interaction layer is represented by an agent interface that provides interaction between a human (user) and his/her counterpart agent. Figure 4 shows the interface for communication exchanges between the user and his/her agent. A user inserts the desired keywords based on the required papers and assigns the threshold for each keyword. Subsequent to the input data being passed to his/her counterpart agent via the interface, the agent takes over and performs its tasks.

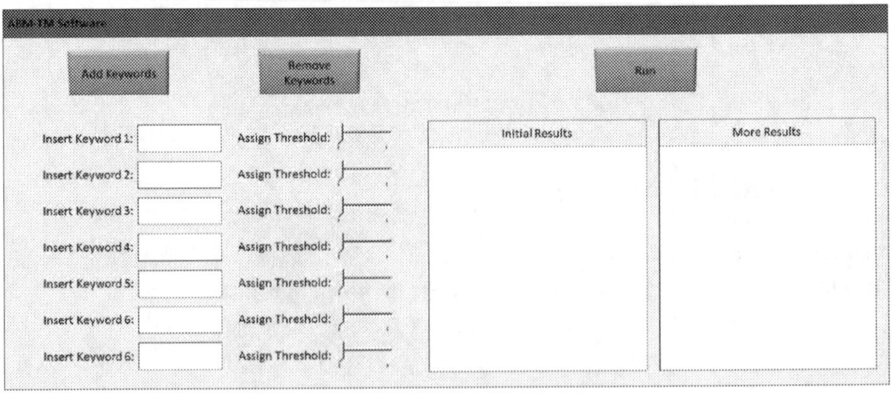

Fig. 4. A Proposed Interface

4.2   The Knowledge Sharing Layer

The knowledge sharing layer is represented by the Global Knowledge Base (GKB). The GKB data is structured based on Case-based Reasoning (CBR) structure. The knowledge within this knowledge base is structured as cases. Each case represents a paper with a specific data. If C is a case, and GKB is the union of all cases, then,

$$GKB = \bigcup_{n=1}^{k} C_k \ where \ k \geq 1$$

To recognize a paper, each case data consist of five or more variables. For example, if C is a case; and v is a variable; then,

$$C = (v_1, v_2, v_3, \dots \dots v_k) \qquad where \ k \geq 5$$

Then,

$$C = (v_{n=5}^{k})$$

Table 1 illustrates the five mandatory variables and the optional variables.

Table 1. Mandatory and Optional Variables

The five mandatory variables	The optional variables
V_1: Article Title V_2: Publisher Title (Journal, Conference, Book Chapters,…) V_3: First Author Name V_4: Keyword 1 V_5: Keyword Strength 1	V_6: Keyword 2 V_7: Keyword Strength 2 V_8: Keyword 3 V_9: Keyword Strength 3

From Table 1, we define C based on the proposed variables as follows:

C = (ArticleTitle, PublisherTilte, FirstAuther, (Keyword1, Strength), (Keyword2, Strength)…)

4.3   The Extraction Layer

We propose a Text Mining Algorithm (TMA) deployed in an agent to extract the potential context of a paper. Due to space limitation, we assume an example to represent an abstract to illustrate the algorithm. Assume W is a word, i.e., world, agent…; v is an adverb, i.e., am, are…; p is a preposition, i.e., of, for…; then the assumed abstract and keywords are as follows:

Abstract: W_3 W_7 W_{10} v_2 W_3, W_{10} W_4 p_1 W_5, W_{11} v_5 W_{10}. W_1 W_3 W_6 v_3 W_7 W_8 p_5 "W_{10} W_9" W_8. W_7 W_4 W_{10} v_3 W_{12} (W_6; W_{11}) p_2 W_9 W_{10}. W_5 W_8 W_6 p_2 W_{10}, W_3 W_{11} W_1 W_8. W_2 W_3 W_4, W_8 v_3 W_7 W_{12}, W_{10} W_3 p_4 W_9.

Keywords: W_3, W_{10}, W_8, W_7, W_9

In this example, a user inputs the keywords and threshold as follow:

Keyword	Threshold
W_{10}	80%
W_3	60%
W_8	60%

We assume that the agent search online and found the above mock abstract's keywords that match the user's input keywords. The agent runs the TMA to mine and extract and eventually identify if the paper is or is not suitable.

1   Data processing: It is the first stage to have appropriate data format for mining. The process steps are:

- Punctuations filtering: is the process of removing all the punctuation marks i.e. (,), (;) from clauses to be cleared for formatting. Applying this step to the mock abstract, the following is obtained:

    $W_3 \ W_7 \ W_{10} \ v_2 \ W_3 \ W_{10} \ W_4 \ p_1 \ W_5 \ W_{11} \ v_5 \ W_{10} \ W_1 \ W_3 \ W_6 \ v_3 \ W_7 \ W_8 \ p_5 \ W_{10} \ W_9 \ W_8$
    $W_7 \ W_4 \ W_{10} \ v3 \ W_{12} \ W_6 \ W_{11} \ p_2 \ W_9 \ W_{10} \ W_5 \ W_8 \ W_6 \ p_2 \ W_{10} \ W_3 \ W_{11} \ W_1 \ W_8 \ W_2 \ W_3$
    $W_4 \ W_8 \ v_3 \ W_7 \ W_{12} \ W_{10} \ W_3 \ p_4 \ W_9$

- Data formatting: This process restructures the data into a form that is mineable. To do so, this step tokenizes the clauses by replacing spaces with a comma. Applying this step to the mock abstract, the following is obtained:

    $W_3, \ W_7, \ W_{10}, \ v_2, \ W_3, \ W_{10}, \ W_4, \ p_1, \ W_5, \ W_{11}, \ v_5, \ W_{10}, \ W_1, \ W_3, \ W_6, \ v3, \ W_7, \ W_8, \ p_5,$
    $W_{10}, \ W_9, \ W_8, \ W_7, \ W_4, \ W_{10}, \ v_3, \ W_{12}, \ W_6, \ W_{11}, \ p_2, \ W_9, \ W_{10}, \ W_5, \ W_8, \ W_6, \ p_2, \ W_{10},$
    $W_3, \ W_{11}, \ W_1, \ W_8, \ W_2, \ W_3, \ W_4, \ W_8, \ v_3, \ W_7, \ W_{12}, \ W_{10}, \ W_3, \ p_4, \ W_9$

- Data filtering: It is the process of removing repeated instances, articles (a, an, the), personal pronouns (he, she, etc.), adverbs (am, is, etc.), prepositions (of, in, etc.) of the formatted data to get a unique set, which forms the candidate keywords. Applying this step to the mock abstract, the following is obtained:

    $W_3, \ W_7, \ W_{10}, \ W_4, \ W_5, \ W_{11}, \ W_1, \ W_6, \ W_8, \ W_9, \ W_{12}, \ W_2$

- Candidate keywords: This is an instance of a keyword that is discovered as a consequence of data formatting and filtering. The candidate keywords set contains all keywords including the potential and weak keywords. According to the mock abstract, the formatted and filtered instance is as obtained earlier.

    Candidate Keywords $= W_3, \ W_7, \ W_{10}, \ W_4, \ W_5, \ W_{11}, \ W_1, \ W_6, \ W_8, \ W_9, \ W_{12}, \ W_2$

2   Identify and verify the Identity Keyword: As mentioned earlier, the identity keyword represents the highest repeated keyword from the candidate keywords set. According to the proposed abstract,

- Determine the frequency of candidate keywords, from the abstract:

W_3	= 6	W_1	= 2	
W_7	= 4	W_6	= 3	
W_{10}	= 8	W_8	= 5	
W_4	= 3	W_9	= 3	
W_5	= 2	W_{12}	= 2	
W_{11}	= 3	W_2	= 1	

- Determine the highest number of repeated candidates keyword:

$W_{10} = 8$
If W_{10} belongs to the set of paper keywords, then it is considered as the Identity keyword, otherwise it is not.
According to the keywords of the mock abstract, W_{10} is one of the keywords, then
Identity Keyword = ($W_{10} = 8$)

3   Identify a Keyword Strength Percentage: Divide 100 by the identity keyword frequency to get the percentage of strength for a candidate keyword:

Keyword Strength Percentage = 100/ Identity Keyword Frequency
Keyword Strength Percentage = 100/8 => Keyword Strength Percentage = 12.5%

4   Calculate the strength of each keyword by multiplying each keyword frequency by the Keyword Strength Percentage.

W_3	= 6 * 12.5 = 75%	W_1	= 2 * 12.5 = 25%
W_7	= 4 * 12.5 = 50%	W_6	= 3 * 12.5 = 37.5%
W_{10}	= 8 * 12.5 = 100%	W_8	= 5 * 12.5 = 62.5%
W_4	= 3 * 12.5 = 37.5%	W_9	= 3 * 12.5 = 37.5%
W_5	= 2 * 12.5 = 25%	W_{12}	= 2 * 12.5 = 25%
W_{11}	= 3 * 12.5 = 37.5%	W_2	= 1 * 12.5 = 12.5%

5   Extract Potential Keywords: This step is based on the user input. When the user input the desired keywords and thresholds, the agent uses them as a reference to select the suitable paper. When the assigned thresholds are less than or equal to the observed keyword strengths, then the paper is relevant to the user. Otherwise it is not.

The paper's mock abstract is relevant to the user because each assigned threshold is less than the keyword's strength as follows:

Keyword	Threshold		Keyword Strength
W10	80%	<	100%
W3	60%	<	75%
W8	60%	<	62.5%

5   Conclusion and Further Work

This paper presents a text mining algorithm deployed in a software agent to extract potential context of papers published in the WWW and subsequently determine which papers are potentially relevant to a user's need. The user only input the keywords and assigns a threshold value for each keyword. The agent searches the relevant documents in the GKB and WWW and runs the Text Mining Algorithm to find papers that match with those keywords with strength greater than or equal to the assigned threshold for each keyword.

The text mining algorithm mines keywords in abstracts and identifies the paper's potential context. The algorithm entails operations of data cleaning, formatting, filtering, and identifying the candidate keywords. It subsequently, calculates the strength of each keyword and identifies the potential context based on the assigned threshold value.

Since this work is in its early stage, it only presents the conceptual underpinnings of pertinent issues in search and does not present the experimental results. Such outcome will be presented in our future work. In our future work, we shall also implement the ABM-TM Framework on real data and add some features that allow users to evaluate the search outcome.

References

1. Hotho, A., Nürnberger, A., Paaß, G.: A Brief Survey of Text Mining. LDV Forum 20(1), 19–26 (2005)
2. Feldman, R., Dagan, I.: Kdt - knowledge discovery in texts. In: Proc. of the First Int. Conf. on Knowledge Discovery (KDD), pp. 112–117 (1995)
3. Kotsiantis, S., Kanellopoulos, D.: Association Rules Mining: A Recent Overview. International Transactions on Computer Science and Engineering 32(1), 71–82 (2006)
4. Ogunde, A., Follorunso, O., Sodiiya, A., Oguntuase, J., Ogunlleye, G.: Improved cost models for agent-based association rule mining in distributed database. Anale. Seria Informatica. IX (1), 231–250 (2011)
5. Symeonidis, A.L., Mitkas, P.A.: Agent Intelligence Through Data Mining. In: The 17th European Conference on Machine Learning and The 10th European Conference on Principles and Practice of Knowledge Discovery in Databases (2006)
6. Chiwara, M., Al-Ayyoub, M., Hossain, M.S., Gupta, R.: Data Mining Concepts and Techniques Association Rule Mining, State University of New York, CSE 634, Chapter 8 (2006)
7. Poelmans, J., Ignatov, D.I., Viaene, S., Dedene, G., Kuznetsov, S.O.: Text mining scientific papers: A survey on FCA-based information retrieval research. In: Perner, P. (ed.) ICDM 2012. LNCS, vol. 7377, pp. 273–287. Springer, Heidelberg (2012)

8. Liu, X.: Learning from multi-view data: clustering algorithm and text mining application. Katholieke Universiteit Leuven, Leuven, Belgium (2011)

9. Aase, K.: Text Mining of News Articles for. Stock Price Predictions. Trondheim, Master's thesis. Trondheim (June 2011)

10. Nahm, U.Y., Mooney, R.J.: Text Mining with Information Extraction. In: Proceedings of the AAAI 2002 Spring Symposium on Mining Answers from Texts and Knowledge Bases, pp. 60–67. Stanford, CA (March 2002)

11. Zhong, N., Li, Y., Wu, S.T.: Effective pattern discovery for text mining. IEEE Transactions on Knowledge and Data Engineering (2011)

12. Jusoh, S., Alfawareh, H.M.: Agent-based knowledge mining architecture. In: Proceedings of the 2009 International Conference on Computer Engineering and Applications, IACSIT, pp. 602–606. World Academic Union, Manila (June 2009)

13. Lai, K.K., Yu, L., Wang, S.-Y.: Multi-agent web text mining on the grid for enterprise decision support. In: Shen, H.T., Li, J., Li, M., Ni, J., Wang, W. (eds.) APWeb Workshops 2006. LNCS, vol. 3842, pp. 540–544. Springer, Heidelberg (2006)

A Method for Response Integration
in Federated Data Warehouses

Rafał Kern[1], Grzegorz Dobrowolski[2], and Ngoc Thanh Nguyen[1]

[1] Wroclaw University of Technology, Wroclaw, Poland
[2] AGH University of Science and Technology, Krakow, Poland
{Rafal.Kern,Ngoc-Thanh.Nguyen}@pwr.edu.pl,
Grzela@agh.edu.pl

Abstract. In this paper a data integration method for data warehouse federations has been proposed. It often happens that for a query different component warehouses can generate different (even inconsistent) responses. Thus it is needed to integrate them creating a common response of the federation. In this paper the most common types of responses are discussed and an algorithm for response integration has been worked out.. We also consider a component data warehouse to be described by some factors which may be used as additional input for the integration method.

Keywords: data integration, data warehouses, federation.

1 Introduction

Data warehouses federation is widely used to obtain integrated data from many distributed data sources ([9], [10]). The subject of data integration is described in many works ([8], [13]). The federation as a fully functional tool must be able to perform three actions:

- Integrating various schemas of existing components,
- decomposing the user query into forms which are understandable for each component,
- integrating the partial responses from component data warehouses.

In this paper the last action will be discussed. Luckily, the architecture of federation and the specific kind of data allow us to make some assumptions in designing the data integration process.

There are many types of queries which may be sent to data warehouses: annual income, sale trends, top products, the most profitable customer etc. That information may be used to support some decision process in different markets. Especially in geographical federation the integration methods should comply facts like different currencies, metric systems etc. This data should be stored in data dictionary ([2],[4]). Responses for many of those queries may be integrated using different integration methods. In this paper some different situations for geographical federation are described.

© Springer International Publishing Switzerland 2015
D. Camacho et al. (eds.), *New Trends in Computational Collective Intelligence*,
Studies in Computational Intelligence 572, DOI: 10.1007/978-3-319-10774-5_6

The other problem with data integration received from distributed data sources is that in most cases than should not be treat the same way. In other word, some of them should influence the final result according to some factors. In next sections some examples of factors are described.

The paper is organized as follows: Section 2 contains analysis of related works, Section 3 contains necessary definitions and notions. In Section 4 some data warehouses factors are described and in Section 5 a list of responses types is provided. Section 6 contains the idea of integration with some examples. Summary and some propositions for future works are presented in Section 7.

2   Related Works

In many papers the data integration in federated data warehouses is described as a very complex process. It is a natural consequence of query decompositions and processing sub-queries in different data sources. Both of these processes are necessary in a successful federation tool [1], [4], [6], [14], [15]. In our approach both of them take place in federation level which must hide the data sources heterogeneity and dispersion. The process usually uses the mappings between global and local schemas. Beyond these two elements some additional data may be stored in federation level, e.g. some metadata gained during ETL process or some totally external data given by an expert. The federation seems to be a good and intuitive way for solving this problem in distributed data sources.

In [8] an analysis of data integration was discussed with many challenges and possibilities. The idea of Data Integration System proposed there is often developed into part of FDW Management System. The DIS contains three parts: global schema, source schema and mappings between them. It also introduces the 'Local As View' and 'Global As View' principles. Mentioned idea was used in some other approaches (e.g. [2]) where semantics was applied to improve the matching between proper measures. To the final result leads a sequence of splitting and merging proper measures and facts but the final decision, which operator should be chosen always should be made by a human. Another solution is described in [14] where the integration problem is solved by creating virtual SQL view. With global schema described as the largest common schema, mappings between measures and dimensions and UNION operator the FDW system is able to produce satisfying result. However - as the author mentioned - some performance and interpretation problems may arise.

In general, two approaches toward data integration are available [7]. The first one minimize individual dissatisfaction which is based on distances between components answers. The second one minimizes the global dissatisfaction represented as a sum of distances between answers from each component and federation. As this paper describes integration in federation system which main task is to provide some global scope on distributes data, the second approach seems to be a better choice. The integration should be performed in qualitative level [16]. Besides that, it should accept some additional input like weights defined by user[10].

In our approach we continue development of the idea proposed in [4]. We propose different integration methods for different response types. Some types may be integrated in more than one way and we believe that the end user should have the final choice because only him knows best his own needs. To reach this goal we introduce two steps approach where the integration phase is preceded by normalization. In the normalization phase data is transformed into one common units system and grouped by matching attributes.

Inconsistency very often happens during the integration processes. Different data warehouses can contain inconsistent data on the same subject, therefore the responses for the same query can be in conflict. Thus conflict resolution is often needed. In works [18], [19] and [20] some methods for processing inconsistency of knowledge have been proposed. Paraconsistent logics have also been proved to be useful for this kinds of processes [21]. Federations of data warehouses can be considered as an effective solution for big data processing problems [17].

3   Basic Notions

According to notion described in [1] and [2] we define data warehouses as

$$H_i = (D_0^i, D_1^i, D_2^i, \dots, D_{\alpha i}^i)$$

where i is the index of data warehouse in federation, D_0^i is the fact table, D_j^i is a dimension for $j = 1, 2, \dots, \alpha i$.

The federation $\hat{F}$ built on data warehouses $H_1, H_2, \dots, H_n$ is denoted as

$$\hat{F} = (F, H, U, q, l)$$

where:

  $\hat{F}$ – federation
  F – federation schema
  H - set of data warehouses
  U – user/application interface
  q – query decomposition procedure
  l – answers integration procedure
  The federation schema is denoted as:

$$F = (D_0, D_1, D_2, \dots, D_m)$$

where D_0 is the fact table and D_j is a dimension for $j = 1, 2, \dots, m$.

Every component data warehouse H_i may generate an answer, which is sent back to federation layer. It is denoted as:

$$R_i = (S_i, data_i, \varphi_i)$$

where:
 S_i is a cube schema slice (subset of F) including calculated measures
 $data_i$ is a set of raw records:

$$data_i = \{(key^{i,1}, value^{i,1}), (key^{i,2}, value^{i,2}), \dots, (key^{i,t_i}, value^{i,t_i})\}$$

where

- t_i is the number of records sent in answer R_i,
- $(key^{i,j})$ is a set of key values (attributes)
$$(key^{i,j}) = (k_1^{i,j}, k_2^{i,j}, \dots, k_p^{i,j}),$$
- $k_s^{i,j}$ is a single value of an attribute. s – order number in every row. Sometimes $k_s^{i,j}$ may be null-valued , because $key^{i,j} = \emptyset$. It means that answer contains only one row.
- $k_s^{i,j}$ is value of each attribute stored in $(d_1^i \cup d_2^i \cup \dots \cup d_k^i)$.

During recent works one mayor change occurred. The $value^{i,j}$ cannot be a simple one. In some cases it should be a set of values. Sometimes an ordered one. Therefore in further parts of this paper we will use following notion:

$$(key^{i,j}) = (k_1^{i,j}, k_2^{i,j}, \dots, k_p^{i,j})$$
$$(value^{i,j}) = (v_1^{i,j}, v_2^{i,j}, \dots, v_m^{i,j})$$

where:

p - number of attributes in one record

m - number of measures in one record

Let φ_i be a set of projection functions which may be used for normalization and shows the importance of every single data warehouse. It contains functions transforming attributes and measures. Set φ_i shows how to interpret them (e.g. how to compare prices in different currencies systems). According to $data_i$ description provided above, $\varphi_i^{(value)}$ sometimes must contain more than one function

$$\varphi_i = \left(\varphi_i^{(key)}, \varphi_i^{(value)} \right)$$

$\varphi_i^{(key)} = (\varphi_i^{k_1}, \varphi_i^{k_2}, \dots, \varphi_i^{k_p})$ – ordered set of functions modifying attribute values

$\varphi_i^{(value)} = (\varphi_i^{v_1}, \varphi_i^{v_2}, \dots, \varphi_i^{v_m})$ - ordered set of functions transforming measures values in every record.

The projection functions must be provided by the expert or worked out during schema integration. It should be stored in data dictionary in federation layer. Usually it has a very simple form (e.g. linear function), however, sometimes it may be very complicated. When applying more sophisticated types we must remember that it may strongly affect the performance of federation in big data sets.

4   Data Warehouses Metrics

For the need of integration processes in data warehouse federations we use the following metrics:

- **Power** - the ratio between amount of data in DW facts table and the whole federation. Directly is described as the number of records in facts table. The more data is stored, the more credible this data warehouse is.

$$p = \frac{Card(H_i)}{Card(F)}$$

$p \in <0,1>$ - power always have 0 do 1.
$Card(H_i)$ - number of facts in specific data warehouse
$Card(F)$ - number of facts in whole federation

- **Coverage** - warehouse schema coverage by the federations schema. How many of data warehouse attributes and measures are used in the final federation schema.

$$c = \frac{AC(H_i) + MC(H_i)}{AC(F) + MC(F)}$$

$AC(H_i)$ - number of all dimensions attributes which are related with attributes in federation schema
$MC(H_i)$ - number of all measures which are related with measures in federation schema
$AC(F)$ - number of attributes in federation schema
$MC(F)$ - number of measures in federation schema
- **Diversity** - domain diversity (set of different values stored in dimensions), in particular (by geographical localization and time)

$$d_D = \frac{card(D_j^i)}{card(D_j)}$$

$card(D_j^i)$ - number of records in data warehouse dimension D_j^i
$card(D_j)$- number of records in all data warehouses dimensions related with D_j

5   Types of Responses

The following types of responses are defined:
- **Scalar** - the simplest response type. Answers the question like "*What was the last year annual income?*". A query for that kind of response usually contains some aggregation functions which seems to be a good tip how to integrate the partial responses. According to definition given in section 3 it contains no attributes.
- **Rank** - the query contains clauses ORDER BY and WHERE. The attributes $(k_1^{i,j}, k_2^{i,j}, ..., k_p^{i,j})$ are less important than the order.

SELECT * FROM Table ORDER BY col1, col2, col3;

This kind of sorting makes more important the columns placed closer to ORDER BY clause. Usually occurs with LIMIT key word. During integration sometimes it is necessary to calculate only the not null values given in response (e.g. when weighted average is used).
- **Tendency** - the record are sorted by the time dimension. Usually used to make some prediction about the future in global scope. It may be called a specific kind of

Rank with sorting by one very specific dimension. The time dimension is a very characteristic one which makes us think that it deserves special treatment. At least on U_{k_x} must refer do the time dimension.

- **Structural Pattern** - answers the question like: "*Which product are usually bought together in one transaction?*". The order of $(v_1^{i,j}, v_2^{i,j}, ..., v_m^{i,j})$ is not important. The most important is the fact that some certain items occurred in one set. Of course it seems to be obvious that, the number of values (m) should be greater than 1.

- **Behavioral Pattern** - answers the question like: "*Which sequences of shopping activities can be noticed?*". E.g. some customers firstly has bought product A, later - product B and finally - product C. This type may be used in many recommendation systems because it allows to predict the next step of the customer. In this kind of response the most important is the order of each action. Therefore, during integration the time dimension should be treated as the most important. It differs from Tendency because it is focused on average customer instead of whole population.

- **Set** - any other kind of response. In this type it is possible to find some other types. Depending on the aggregation level the attributes used for grouping should be faced to their domain (e.g. if the results are grouped by country/region we should check how many countries are in this geographical dimension).

- **Simple set** - the data comes only from the fact table.

- **Complex set** -the data comes from facts table and dimensions.

6   Algorithm for Response Integration

After receiving responses from different data warehouses we have a situation where different agents give different information about similar things. Usually we can treat it as a conflict which could be resolved in various ways. In general, it is necessary to give a response which describes the partial responses from the global scope in best way.

The integration runs in the federation layer described widely in [4]. In general it runs in few steps:

1. Gather all the partial responses
2. Using φ_i normalize the partial responses into common types
3. Integrate the data respecting attributes compatibility. If possible, use aggregation methods:
    a. COUNT - sum or average, may be weighted
    b. SUM - sum
    c. AVG - average, may be weighted
    d. MIN - minimum value
    e. MAX - maximum value
4. If the query contains ORDER BY values sort the result by given parameter.
5. If the query contains LIMIT clause use only declared number of records.

For simplification and better visualization we assume that m = 1.
In more formal way we have:

- 1st phase:

R* = {} - responses matrix
for (i=1; i<= L; i++)
begin

    for (j = 1; j <= t_i; j++)
    begin

$$R*[\varphi_i^{key}(key^{i,j})][i] = \varphi_i^{value}(value^{i,j})$$

    end

end

this is the end of the normalization phase. As a result we get matrix as below:

	K1	K2	K3	K4	K5
H_1	$\varphi_1^{value}(value^{1,1})$	$\varphi_1^{value}(value^{1,2})$		$\varphi_1^{value}(value^{1,4})$	
H_2	$\varphi_2^{value}(value^{2,1})$		$\varphi_2^{value}(value^{2,3})$	$\varphi_2^{value}(value^{2,4})$	
H_3		$\varphi_3^{value}(value^{3,2})$	$\varphi_3^{value}(value^{3,3})$		$\varphi_3^{value}(value^{3,5})$

- 2nd phase:

    for (k in keys)
    begin

        R[k] = I(R*[k])

    end

where I is the integration function for values with the same keys.

According to architecture introduced in [5] the $data_i$ comes from component data warehouse, φ_i are stored in federation layer and S_i is attached to each query. It is indispensable to allow the federation to interpret the result and integrate with other partial records. The federation 'knows' how attributes from native schemas of each component are matched with the global schema and S_i is the minimum set of attributes to understand the data obtained from component layer.

Example 1:

Responses type - rank:

Get 10 countries with the largest number of transactions.

```
SELECT  C.idCountry
FROM Sales AS S, Customers AS C
WHERE S.IdCustomer = C.IdCustomer
ORDER BY Count(S.IdSale)
LIMIT 10
```

It is possible that none measure is assigned to attributes. The attributes itself contains large dose of information which connected to its position in response allows to form an answer. We treat the data warehouse as a black box. We know what is given on the input and what we receive on output. We have no information about any of the operations inside.

We have answers from 3 component data warehouses.

$R_1 = \{((DE),null), ((NL),null), ((FR), null), ((AT), null), ((IT), null)\}$
$R_2 = \{((CZ),null), ((SK),null), ((HU), null), ((DE), null), ((AT), null)\}$
$R_3 = \{((DE),null), ((BE),null), ((NL), null\}$

For each record we use normalization procedure which in this kind of response should give some wages instead of values ($value^{i,j}$).

$$\varphi_i^{value}(value^{i,j}) = t^i - j$$

t^i - number of items in one response
$t^1 = 5, t^2 = 5, t^3 = 3$
$R^*[\varphi_i^{key}(key^{i,j})][i] = \varphi_i^{value}(value^{i,j})$

As a result we have:

H_i/keys	DE	NL	FR	AT	IT	CZ	SK	HU	BE
H_1	5	4	3	2	1				
H_2	2			1		5	4	3	
H_3	3	1							2

Proceeding into next step (integration using attributes values) we have:

	DE	NL	FR	AT	IT	CZ	SK	HU	BE
R	10	5	3	3	1	5	4	3	2

So $I(R^*[k]) = \sum_{i=0}^{L} R^*[k][i]$

But we may take into account the cardinality in every response.

	DE	NL	FR	AT	IT	CZ	SK	HU	BE
R	3,33	2,5	3	1,5	1	5	4	3	2

After sorting we have:

	CZ	SK	HU	FR	DE	NL	BE	AT	IT
R	5	4	3	3	3,33	2,5	2	1,5	1

Example 2:
Responses type - set. Get top sets of products bought together.

As a results we get:
$R_1 = \{((P1, P7, P11, P25, P31), null)\}$
$R_2 = \{((P7, P9, P11, P24, P26), null)\}$
$R_3 = \{((P7, P12, P14, P15, P21), null)\}$

In this example the result does not contain any real measure. Products names are in fact values of attributes. However, sum(quantity) may be considered as a calculated measure. It is easy to notice, than now the $value^{i,j}$ is not a simple one, but a set of values. Using integration method which takes into account each products cardinality and sorting by the cardinality the final set look as below (cardinality in brackets):

R = {((P7 [3], P11 [2], P1 [1], P25 [1]), P31 [1]), null)}

The first two product positions are absolutely clear, but position 3-5 are not. This is a problem may be solved by consensus theory ([11], [12]) or applying some additional factors that allow to treat some data warehouses as "more important".

Example 3:

Get total income in first 4 month:

SELECT SUM(`priceNetto` * `quantity`) AS total FROM `sales` , `date`
WHERE sales.`idCreateDate` = `date`.idDate
GROUP BY `date`.`year` , `date`.`month`
ORDER BY `idCreateDate`

H_i/keys	(2013, 1)	(2013, 2)	(2013, 3)	(2013, 4)
H_1	17768	239789	272316	236067
H_2	70643	1703376	2192232	1932092
H_3	8716	161063	162217	176301

In this example H_2 stores prices in USD but the others - in EUR. The final result should be in EUR too. So we need to introduce $\varphi_i^{value}(value^{i,j}) = CEIL(value^{i,j} * 0{,}723)$ (USD/EUR = 0.723)

After normalization we have:

	(2013, 1)	(2013, 2)	(2013, 3)	(2013, 4)
H_1	17768	239789	272316	236067
H_2	51075	1231541	1584984	1396903
H_3	8716	161063	162217	176301

And after integration using AVG function:

	(2013, 1)	(2013, 2)	(2013, 3)	(2013, 4)
R	25853	544131	673172,3	603090,3

As long as we must deal with numerical data the integration problems are rather easy to solve. More complicated situation appears in structural or behavioral patterns. The φ_i functions should take into consideration metrics described in previous sections. The simplest form of such one is weighted average using power metric

7 Conclusions

In this paper the idea from previous works ([4], [5]) has been developed. More detailed description of the two-steps responses integration method has been provided. It is a challenging task to automatically apply the best integration method without good description of user requirements. That is why for now we suggest that the user should have the ability to decide which integration method will he use. Using mentioned metrics in the projection methods in case of non-numerical responses looks promising after initial research. The exact form of this method should be worked out in further works. In addition, an application of the method to real-life problem should be considered. One of the promising problems can be a digitial knowledge database developed for Polish Government Protection Bureau. Not only will this verify the proposed method but also allow for a comparison with the already utilized approaches described in [22].

Acknowledgement. The research reported in the paper was partially supported by the Grant No.\ DOBR-BIO4/060/13423/2013 from the Polish National Centre for Research and Development.

References

1. Akinde, M.O., et al.: Efficient OLAP Query Processing in Distributed Data Warehouses. Inf. Syst. 28(1-2), 111–135 (2003)
2. Berger, S., Schrefl, M.: From Federated Databases to a Federated Data Warehouse System. In: Proceedings of the 41st Hawaii International Conference on System Sciences. IEEE (2008)
3. Jindal, R., Acharya, A.: Federated Data Warehouse Architecture. White paper, Wipro Technologies (2003)
4. Kern, R., Stolarczyk, T., Nguyen, N.T.: A Formal Framework for Query Decomposition and Knowledge Integration in Data Warehouse Federations (2013)
5. Kern, R., Ryk, K., Nguyen, N.T.: A Framework for Building Logical Schema and Query Decomposition in Data Warehouse Federations (2011)
6. Kimball, R., Wiley, J.: The Data Warehouse Toolkit: Practical Techniques for Building Dimensional Data Warehouses. John Wiley & Sons (1996)
7. Konieczny, S., Pino-Pérez, R.: On the Logic of Merging. In: Proceedings of the sixth International Conference on Principles of Knowledge Representation and Reasoning (KR 1998), pp. 488–498 (1998)
8. Lenzerini, M.: Data integration: A theoretical perspective. In: Proc. PODS, pp. 233–246 (2002)
9. Maleszka, M., Mianowska, B., Nguyen, N.T.: A Framework for Data Warehouse Federations Building (2012)
10. Motro, A., Anokhin, P.: Data integration: Inconsistency Detection and Resolution Based on Source Properties. Information Fusion 7(2), 176–196 (2006)
11. Nguyen, N.T.: A Method for Ontology Conflict Resolution and Integration on Relation Level. Cybernetics and Systems 38(8), 781–797 (2007)
12. Nguyen, N.T.: Advanced Information and Knowledge Processing. Springer-Verlag, London (2008)
13. Rizzi, S.: Collaborative Business Intelligence. In: Aufaure, M.-A., Zimányi, E. (eds.) eBISS 2011. LNBIP, vol. 96, pp. 186–205. Springer, Heidelberg (2012)
14. Schneider, M.: Integrated vision of federated data warehouses. In: CEUR-WS (DisWeb 2006), Proceedings of International Workshop on Data Integration and Semantic Web, vol. 238, pp. 336–347 (2006)
15. Waddington, D.: An Architected Approach to Integrated Information. Kalido white paper (2004)

16. Wallgrün, J.O., Dylla, F.: Spatial Data integration with Qualitative Integrity Constraints. In: Purves, R., Weibel, R. (eds.) Online Proceedings of the 6th International Conference on Geographic Information Science. GIScience (2010)
17. Vossen, G.: Big Data As The New Enabler in Business and Other Intelligence. Vietnam Journal of Computer Science 1(1), 3–14 (2014)
18. Duong, T.H., Nguyen, N.T., Jo, G.S.: A Method for Integration of WordNet-based Ontologies Using Distance Measures. In: Lovrek, I., Howlett, R.J., Jain, L.C. (eds.) KES 2008, Part I. LNCS (LNAI), vol. 5177, pp. 210–219. Springer, Heidelberg (2008)
19. Duong, T.H., Nguyen, N.T., Jo, G.S.: Constructing and Mining: A Semantic-Based Academic Social Network. Journal of Intelligent & Fuzzy Systems 21(3), 197–207 (2010)
20. Nguyen, N.T.: Metody wyboru consensusu i ich zastosowanie w rozwiązywaniu konfliktów w systemach rozproszonych. Wroclaw University of Technology Press (2002)
21. Nakamatsu, K., Abe, J.M.: The paraconsistent process order control method. Vietnam Journal of Computer Science 1(1), 29–37 (2014)
22. Dajda, J., Debski, R., Kisiel-Dorohinicki, R., Piętak, K.: Multi-Domain Data Integration for Criminal Intelligence. In: Gruca, A., Czachórski, T., Kozielski, S. (eds.) Proc. of ICMMI 2013. AISC, vol. 242, pp. 345–352. Springer, Heidelberg (2013)

Part II
Data Mining Methods and Applications

An Asymmetric Weighting Schema
for Collaborative Filtering

Parivash Pirasteh, Jason J. Jung*, and Dosam Hwang

Department of Computer Engineering,
Yeungnam University, Korea
{parivash63,j2jung,dosamhwang}@gmail.com

Abstract. Several key applications like recommender systems need to determine the similarities between users or items. These similarities play an important role in many tasks, such as discovering users with common interests or items with common properties. Most of the traditional methods are symmetric which means that they always assign equal similarity to each user even when one user rating profile is completely like the other but not conversely. In this paper we combine the traditional methods with an asymmetric weighting schema to distinguish between the similarities of two users. Several experiments have been performed to compare the performance of proposed method with traditional similarity measures.

Keywords: Collaborative filtering, Item similarity, Recommender systems, User similarity.

1 Introduction

Systems with content-based recommendation implementation approach analyze a set of documents and/or descriptions of items previously rated by a user, and build a model of user interests based on the features of the objects rated by that user [1]. The user profiles include additional information relating to the user or associated with the ratings given to an item by the user.

Collaborative Filtering (CF) is complementary to content-based filtering. It aims at learning predictive models of user preferences, interests or behavior from community data which is a database of available user preferences [2]. CF analyzes relationships between users and inter-dependencies among products to identify new user-item associations. CF techniques are more often implemented than content filtering and often result in better predictive performance. The main reason is that they require no domain knowledge and avoid the need for extensive data collection [3]. Instead, they are based on users preferences for items, that can carry a more general meaning than is contained in an item description. Indeed, viewers generally select a movie to watch based upon more elements than only its genre,actors and director [4].

[5] suggests to interchange the role of items and users. therefore, while user-based algorithm generate predictions based on similarities between users, item-based algorithms generate prediction based on similarities between items. The quality of the CF

---

* Corresponding author.

© Springer International Publishing Switzerland 2015
D. Camacho et al. (eds.), *New Trends in Computational Collective Intelligence,*
Studies in Computational Intelligence 572, DOI: 10.1007/978-3-319-10774-5_7

based recommendation systems mostly depends on estimation of the similarity between active user and other users (in user-based CF) or between active item and other items (in item-based CF). Generally, when comparing two users or two items, only the set of attributes in common are considered. However, by doing so, many users and items may be compared based only on very few attributes, which can lead to false recommendation [4].

In this paper, we suggest to use a new weighting scheme of existing similarity measures, so that users and items that share many attributes are preferred to others that only share a few attributes. Also, our weighting schema distinguish between two users when one user' behavior is quite similar to the other but not conversely. To summarize, this paper, on the study of similarity measures for collaborative filtering, will be structured as follows: an overview of the principal approaches for collaborative filtering is presented in section 2, alongside a study of similarity measures for comparing users and items; the new weighting schema is proposed in section 3,the results of extensive experiments are reported in section 4 in order to compare various similarity measures for collaborative filtering approaches, section 5 concludes the paper and topics for future research are suggested in this section.

2   Collaborative Filtering

Collaborative approaches overcome some of the limitations of content-based ones. For instance, items for which the content is not available or difficult to obtain can still be recommended to users through the feedback of other users. Furthermore, collaborative recommendations are based on user-user similarity, instead of relying on content that may be a bad indicator of quality. Finally, unlike content-based systems, collaborative filtering ones can recommend items with very different content, as long as other users have already shown interest for these different items [6].

2.1   User-Based Approaches

In user-based approach, the votes of the active user will be predicted based on some partial information regarding the active user and a set of weights calculated from the user database [7]. We assume that the predicted vote of the active user for item i, $P_{a,i}$, is a weighted sum of the votes of the other users. A set of nearest neighbours is selected. And finally, a method for combining the ratings of those neighbours on the target item needs to be chosen. Let $sim(a, u)$ be that similarity measure between users a and u. The predicted rating of item i for the current user a is the weighted sum of the ratings given to item i by the k neighbours of a, according to following formula :

$$P_{a,i} = \bar{r}_a + \frac{sim(a, u).\sum_{u=1}^{k}(r_{u,i} - \bar{r}_u)}{\sum_{u=1}^{k} \mid sim(a, u) \mid} \tag{1}$$

where $\bar{r}_a$ and $\bar{r}_u$ are the average ratings for the user a and user u and $sim(a, u)$ is the similarity between the user a and user u. The summations are over all the users who have rated the item i.

2.2 Item-Based Approaches

While user-based methods rely on the fact that each user belongs to a certain group whose members share similar taste,[5] look at rating given to similar items. Given a similarity measure between items, this approache first define item neighbourhoods. Then the rating of a user on an item is predicted by using the ratings of the user on the neighbours of the target item. The possible choices of the similarity measure $sim(i, j)$ defined between items i and j will be discussed in the next subsection. Then, as for user-based approaches, the item neighbourhood size K is a system parameter that needs to be defined. And given T_i the neighbourhood of item i, two ways for predicting new user ratings can be considered: weighted sum, and weighted sum of deviations from the mean item ratings:

$$p_{ai} = \bar{r}_i + \frac{\sum_{\{j \in S_a \cap T_i\}} sim(i, j) * (r_{aj} - \bar{r}_i)}{\sum_{\{j \in S_a \cap T_i\}} |sim(i, j)|} \tag{2}$$

$\bar{r}_i$ represents here the mean rating on item i.

3 Limitations of Existing Similarity Measure

There are several similarity metrics to calculate similarity, but three commonly used similarity metrics are: Pearson correlation coefficient (PCC) and cosine similarity (COS) and mean square difference (MSD).The formulas are defined as below:

$$PCC(u, v) = \frac{\sum_{p \in I} (r_{u,p} - \bar{r}_u)(r_{v,p} - \bar{r}_v)}{\sqrt{\sum_{p \in I} (r_{u,p} - \bar{r}_u)^2} \cdot \sqrt{\sum_{p \in I} (r_{v,p} - \bar{r}_v)^2}} \tag{3}$$

$$COS(u, v) = \frac{\vec{r_u} \cdot \vec{r_v}}{\| \vec{r}_u \| \cdot \| \vec{r}_v \|} \tag{4}$$

$$MSD(u, v) = \frac{\sum_{p \in |I_u \cap I_v|} (r_{u,p} - r_{v,p})^2}{|I_u \cap I_v|} \tag{5}$$

where I represents the set of common rating items by user u and v. $\bar{r}_u$ and $\bar{r}_v$ are the average rating value of user u and v respectively. $r_{u,p}$ and $r_{v,p}$ denotes the rating of item p by user u and v respectively. MSD computes the similarity between users based on mean difference of common rated items. The users whose difference is greater than a certain threshold, L, are discarded. Thus, the similarity measure based on MSD can be formulated as:

$$sim(u, v) = \frac{L - MSD(u, v)}{L} \tag{6}$$

Although the three popular similarity measures PCC,COS and MSD, have been proved to be successful in many studies, they have some drawbacks. The most important of them in our context is that, these methods assign equal value for the similarity relation between two users, This means, these methods are based on the assumption that $sim(u, v) = sim(v, u)$. Here we use $sim(u, v)$ to show the similarity between user u and user v.

Table 1. An Example of User-Item Matrix

	Item1	Item2	Item3	Item4	Item5	Item6
user1	4		2			
user2	4	1	2	1	1	1
user3		5		3		
user4		1	2			
user5	4				5	5
user6		1	2	2	1	1

Table 1 gives a hypothetical example of user profile including user rating information for different items. From Table 1 we can see that $user_1$ ratings is quite similar to $user_2$ but not vice versa. Traditional methods can not differentiate between these two users with different rating profile. Therefore, the impact that $user_1$ receives in predicting new item's rating from $user_2$ is equal to the impact that $user_2$ receives from $user_1$ for recommending a new movie.

To avoid this contradiction, we use an asymmetric similarity measure, which is defined as the proportion of co-rated items between users, normalized by number of rated items by active user.

$$sim(u, v) = \frac{2 * (|I_u \cap I_v|)^2}{|I_u| * (|I_u| + |I_v|)} \tag{7}$$

where I_u and I_v represent the set of items rated by user u and v, respectively. The similarity metric based on Eq.7 only considers the number of common ratings between two users. Discarding the absolute value of rating leads to very low accurate similarity measurement. For considering the absolute rating in the formula, we propose to combine Eq.7 with other similarity measures, in order to benefit from their respective advantage points. Since it made the similarity measures asymmetric, the new similarity measures called APCC,ACOS and AMSD

4 Experiments

4.1 Dataset

We conduct the experiments using MovieLens dataset. 500 users have been selected randomly. The ratings scale is in the range 1-5 where 1=bad and 5=excellent. We randomly divide the ratings into two disjoint sets: training and test. The training set utilized to calculate predictions using each algorithm. Data from the test set is used to assess the predictive capability of the model.

4.2 Metrics

To evaluate the prediction quality of the proposed methods, we employ Root Mean Squared Error(RMSE) accuracy metrics.

$$RMSE = \sqrt{\frac{1}{T} \sum_{i,j} (R_{i,j} - \overline{R}_{i,j})^2} \tag{8}$$

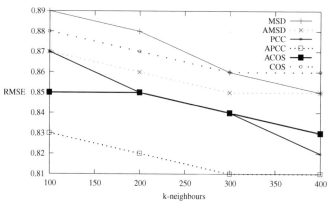

Fig. 1. Comparing RMSE on MovieLens when using item-based approaches with different similarity measures and neighbourhood sizes

Here R_{ij} denotes the rating user i given to item j, $\overline{R}_{i,j}$ denotes the rating user i given to item j as predicted by APCC,ACOS or AMSD methods, and T denotes the number of tested ratings. A smaller value of RMSE signies better prediction quality.

4.3   Results

Fig.1 shows the error rates of item-based approaches depending on the similarity measure used and the neighbourhood size. The optimum value belongs to APCC with 400 neighbours. The asymmetric weighting schema caused progress in performance of similarity measures. Fig.2 shows the same conclusion hold when using user-based approaches. As a result of Fig.1 the APCC outperforms other approaches. The item-based approaches have lower error rate than user-based approaches. It can be interpreted as in

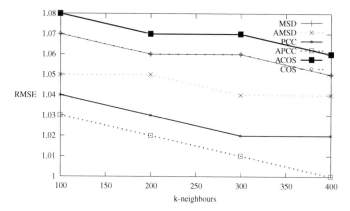

Fig. 2. Comparing RMSE on MovieLens when using user-based approaches with different similarity measures and neighbourhood sizes

subset of data we used the average number of ratings per item is much higher than the average number of ratings per user.

5   Conclusion

In this paper, we have shown that incorporating a weighting schema in user similarity and item similarity approaches can improve the results of CF approaches. This has been confirmed on "movieLens" dataset. The asymmetric characteristic of schema helps to distinguish between the impact that the user(item) has on his neighbor and the impact that the user(item) receives from his neighbor. We have also observed that the error rate of item-based approaches is less than user-based approaches.

Acknowledgment. This work was supported by the BK21+ program of the National Research Foundation (NRF) of Korea.

References

1. Mladenic, D.: Text-learning and related intelligent agents: a survey. IEEE Intelligent Systems 14(4), 44–54 (1999)
2. Hofmann, T.: Latent semantic models for collaborative filtering. ACM Trans. Inf. Syst. 22(1), 89–115 (2004)
3. Koren, Y.: Factorization meets the neighborhood: a multifaceted collaborative filtering model. In: KDD, pp. 426–434. ACM (2008)
4. Candillier, L., Meyer, F., Fessant, F.: Designing specific weighted similarity measures to improve collaborative filtering systems. In: Perner, P. (ed.) ICDM 2008. LNCS (LNAI), vol. 5077, pp. 242–255. Springer, Heidelberg (2008)
5. Sarwar, B.M., Karypis, G., Konstan, J.A., Riedl, J.: Item-based collaborative filtering recommendation algorithms. In: WWW, pp. 285–295. ACM (2001)
6. Ricci, F., Rokach, L., Shapira, B., Kantor, P.B.: A comprehensive survey of neighborhood-based recommendation methods. In: Recommender Systems Handbook, pp. 107–144. Springer, Heidelberg (2011)
7. Breese, J.S., Heckerman, D., Kadie, C.M.: Empirical analysis of predictive algorithms for collaborative filtering. In: UAI, pp. 43–52. Morgan Kaufmann (1998)

Churn Prediction in Telecommunication Industry Using Rough Set Approach

Adnan Amin[1], Saeed Shehzad[2], Changez Khan[1], Imtiaz Ali[1], and Sajid Anwar[1,*]

[1] Institute of Management Sciences Peshawar, Zip Code: 25000, Pakistan
{geoamins,imtiazkmu,sajidanwar.2k}@gmail.com
[2] City University of Science and Technology, Peshawar, Zip Code: 25000, Pakistan
saeedshehzad@gmail.com

Abstract. The Customer churn is a crucial activity in rapidly growing and mature competitive telecommunication sector and is one of the greatest importance for a project manager. Due to the high cost of acquiring new customers, customer churn prediction has emerged as an indispensable part of telecom sectors' strategic decision making and planning process. It is important to forecast customer churn behavior in order to retain those customers that will churn or possible may churn. This study is another attempt which makes use of rough set theory, a rule-based decision making technique, to extract rules for churn prediction. Experiments were performed to explore the performance of four different algorithms (Exhaustive, Genetic, Covering, and LEM2). It is observed that rough set classification based on genetic algorithm, rules generation yields most suitable performance out of the four rules generation algorithms. Moreover, by applying the proposed technique on publicly available dataset, the results show that the proposed technique can fully predict all those customers that will churn or possibly may churn and also provides useful information to strategic decision makers as well.

Keywords: Churn Prediction, Rough Set Theory, Classification.

1   Introduction

Customer churn is one of the mounting issues of today's rapidly growing and competitive telecom sector. The focus of the telecom sector has shifted from acquiring new customer to retaining existing customers because of the associated high cost [1]. The retention of existing customers also leads to improved sales and reduced marketing cost as compared to new customers. These facts have ultimately resulted in customer churn, prediction activity to be an indispensable part of telecom sector's strategic decision making and planning process. Customer retention is one of the main objectives of CRM (customer relationship management). Its importance has led to the development of various tools that support some important tasks in predictive modelling and classification.

---

* Corresponding author.

© Springer International Publishing Switzerland 2015
D. Camacho et al. (eds.), *New Trends in Computational Collective Intelligence*,
Studies in Computational Intelligence 572, DOI: 10.1007/978-3-319-10774-5_8

In recent decades, the organizations are increasingly focusing on long term relationships with their customers and observing a customer's behavior from time to time using various applied knowledge discovery (KDD) techniques [2], [3], [4], [5] to extract hidden relationships between the entities and attributes in a flood of data bank. These facts have attracted many companies to invest in CRM to maintain customers' information. Customer centric approach is very common, particularly in telecommunication sector to predict the customers' behavior based on historical data stored in CRM. Data maintained in such CRM systems can be converted into meaningful information to consider the mounting issue of customer churn to identify the customer's churn activities before the customers are lost which increase the customer strength [6].

Although customer churn prediction modeling has been widely studied in various domains such as financial services, social network services, telecommunication, airlines, online gaming and banking [7]. However, rough set theory application has not been widely studied in customer churn prediction in telecommunication sector. Therefore, this study is approaching to explore the powerful applications of rough set theory for churn prediction in telecommunication sector by constructing more appropriate predictive classifiers that forecast churn prediction based on accumulated knowledge.

The rest of the paper is organized as follows; the next section presents customer churn and related prediction modelling. The preliminary study about rough set theory is explored in section 3 and evaluation methods described in section 4. The evaluation setup and experiments are discussed in section 5 followed by results and comparison in section 6. The paper is concluded in last section 7.

2   Customer Churn and Churn Prediction Modeling

2.1   Customer Churn

Customer churn— Shifting from one service provider to next competitor in the market. It is a key challenge in high competitive markets, which is highly observed in telecommunication sector [5]. Literature reveals the following three types of customer churns [8];

- Active churner (Volunteer): Those customers who want to quit the contract and move to the next provider.
- Passive churner (Non-Volunteer): When the company discontinues the service to a customer.
- Rotational churner (Silent): Those customers who discontinued the contract without prior knowledge of both parties (customer-company).

The first two types of churns can be predicted easily with the help of traditional approaches in term of the Boolean class value, but the third type of churn may exist which is difficult to predict because there may have such type of customers who may possibly churns in near future. It should be the goal of the decision maker and marketers to decrease the churn ratio because it is a well-known phenomenon that existing customers are the most valuable assets for companies as compare to acquiring new one [1].

2.2   Churn Prediction Modelling

Churn prediction has received a tremendous focus from both types of researchers (academia and industry) and have proposed numerous studies for churn prediction in various domains such as financial service [9], Banking Sector [10], [11] Complaint & Repair Services [12], Credit Card Accounts [6], [13], Games [14], [15], Airline Sector [16], Social Networks [17], [18] and Insurance company [19], [20]. Most of the studies have been strongly focused on a few specific factors (e.g: Customer satisfaction& dissatisfaction, loyalty, social influence) relating to customer churn instead of scientifically and empirically investigation and testing of prediction model which encompassing relationships between different constructs such as important churn related variables, switching reasons, service usage, costs and behavior. For instance, in a study [11] classified the customers switching of service with eight general categories. In subsequent work, indicated that the call quality factor is highly influenced on customer churn in the proprietary dataset [21]. Furthermore, surveys based study [22] used a small sample of customer data that may undermine the threat validity and reliability of outcomes.

The literature shows that various machine learning techniques for churn prediction in the telecom industry has been used such as neural network [2],[8],[23],[24], decision table [23],[24],[25], regression analysis [24], SVM [26], [23], Naïve Bayes and Bayesian network [25], evolutionary approach [27] and neuro-fuzzy [28] but conflicts arise when studying results and comparing the conclusion of few of these published works in the area of churn prediction because it is also observed that most of the studies have only evaluated a limited number of traditional machine learning techniques small sample's size [22] and some of them used proprietary dataset. Therefore, the most important problem of which classification technique could use to approach the customer churn prediction in a more appropriate fashion, still remains an open research problem [3]. However, SVM is one of the state-of-the-art technique for classification due to its ability of model nonlinearities but the main drawback is noticed that it generates black-box model [29]. Furthermore, in both studies [8] and [30] stated that artificial neural networks can outperform as compared to other conventional machine learning algorithms.

A benchmarking and empirical study is proposed with the aims to produce further contribution in term of not only achieving suitable accuracy and performance but also extraction of decision rules from hidden existing patterns, based on these rules decision maker can easily adopt new retention policies and improve the overall performance of the organization. In this study, we used rough set theory which has many advantages such as [31]; (1) Finds minimal sets of reduction, (2) Mathematical power to extract hidden patterns from small to large datasets, (3) A straightforward generation and easy interpretation of decision rules. The primarily study about rough set theory is explained in the next section.

3   The Rough Set Theory

Rough Set theory was originally proposed by Pawlak [32] in 1982. A rough set theory has a precise concept of lower and upper approximation and the boundary region. The boundary region separates the lower approximation from upper approximation. Pawlak [33] has defined mathematically rough set approximation and boundaries as; suppose set $X \subseteq U$ and B is an equivalence relation in information system IS= (U, B). Then BL= $\cup$ {Y $\in$ U/ B: Y $\subseteq$ X} is lower approximation and exact member of X while BU = $\cup$ {Y $\in$ U/ B: Y $\cap$ X $\neq\emptyset$} is upper approximation which is possibly a member of X. BR = BL - BU is the boundary region.

3.1   Decision Table

The special case of Information system (IS) is known as a decision table [34]. Formally, information system (IS) is IS = (U, A) where U is a non-empty finite set of instances called the universe and A is a non-empty finite set of attributes or properties i.e $A = \{a_1, a_2, a_3, \ldots a_n\}$ such that a :U$\rightarrow$Va for every a $\in$ A. A decision system is any information system of the form S = (U, C $\cup$ {d}), where C are conditional attributes and d $\notin$ C is the decision attribute. The Union of C and {d} are elements of Set A.

3.2   Indiscernibility, Reduct and Core

The information system IS = (U, A), For any subset of attributes B $\subseteq$ A indiscernibility relation IND (B) is defined as follows: If IND (B) = IND (B-{a}) then a $\in$ B is dispensable otherwise indispensable in B while set B can be called independent if all attributes are indiscernible. If (i, j) $\in$ U $\times$ U belong to IND (B) then we can say that i and j are indiscernible by attributes from B. The reduct's notion can be define by indiscernibility relation. If B $\subset$ A and IND (b) = IND (B). The reduction process is finding more important attributes that preserve discernibility relation with the information. If A be the attribute set of universe set U and B is reduced attribute. Therefore, B is equal of sub-set A and the equation for reduced set is composed as; RED (B) $\subset$ A. The core is the intersection of all reducts. Core (B) = Red (B). Where Red (B) is set of all reducts of B.

3.3   Cut and Discretization

By cut a variable $a_i \in$ IS where IS is an information system, such that Value of a_i is an ordered set and it can be denoted as value c $\in$ Vai. Actually cut mostly appear in the discretization process as pair which determines a split of interval into two disjoint sub-intervals. Discretization is a process of grouping the attribute's data based on the calculated cuts and the continuous variables, converting into discrete attributes [6]. There may exists such unseen object which cannot match with the rules or it can increase computational cost that slow down the machine learning process, so cut & discretization methods are used in order to get high quality of classification [34].

3.4   Rules Generation

The decision rules can be constructed by overlaying the reduct sets over the decision table. It can mathematically express as; $(a_{i1} = v_1) \wedge .....\wedge (a_{ik} = v_k) => d = v_d$, where $1 \leq i_1 < ... <i_k \leq m$, $v_i \in V_{ai}$; Usually the expression represents as; IF C THEN D where C is set of conditions and D is decision value. To extract the decision rules, the following classification algorithms can be used [34]; (i) *Exhaustive Algorithm (GA):* It takes subsets of features incrementally and then returns the reducts of required one. It needs more concentration because it may lead to extensive computations in case of complex and large decision table. It is based on a Boolean reasoning approach [35] (ii) *Genetic Algorithm:* It is based on order-based GA coupled with heuristic and this evolutionary method is presented by [36], [37]. it is used to reduce the computational cost in large and complex decision table.   (iii) *Covering Algorithm:* it is customized implementation of the LEM2 idea and implemented in the RSES covering method. It was introduced by Jerzy Grzymala [38]. (iv) *RSES LEM2 Algorithm:* it is a separate-&-conquer technique paired with lower and upper approximation of rough set theory and it is based on local covering determination of each object from the decision class [38], It is implementation of LEM2 [34].

4   Evaluation Measures

It is nearly impossible to build a perfect classifier or a model that could perfectly characterize all the instances of the test set [39]. To assess the classification results we count the number of True Positive (TP), True Negative (TN), False Positive (FP) and False Negative (FN). The FN value actually belongs to Positive P (e.g. TP + FN = P) but wrongly classified as Negative N (e.g. TN + FP = N) while FP value actually part of N but wrongly classified as P. The following measures were used for the evaluation of proposed classifiers and approaches.

— *Sensitivity:* It measures the fraction of churn customers who are correctly identified as true churn.

$$\text{Sensitivity (Recall)} = \frac{TP}{P} \tag{1}$$

— *Specificity:* It measures the fraction of true non-churns customers, which are correctly identified.

$$\text{Specificity} = \frac{TN}{N} => 1 - \text{Specificity} = \frac{FP}{N} \tag{2}$$

— *Precision:* It is characterized the number of correctly predicted churns over the total number of churns predicted by proposed approach. It can formally express as;

$$\text{Precision} = \frac{TP}{TP + FP} \tag{3}$$

— *Accuracy:* Overall accuracy of the classifier can calculate by the given formula in equation 4.

$$\text{Acc} = \frac{TP + TN}{P + N} \tag{4}$$

— *Misclassification:* Error on the training data is not a good indicator of performance on future data. Different types of errors can be calculated as;

$$\text{Misclassification Error (MisErr)} = 1 - \text{Accuracy} \tag{5}$$

$$\text{Type-I Error} = 1 - \text{Specificity} = \frac{FP}{FP+TN} \tag{6}$$

$$\text{Type-II Error} = 1 - \text{Sensitivity} = \frac{FN}{TP+FN} \tag{7}$$

— *F-Measure:* A composite measure of precision and recall to compute the test's accuracy. It can be interpreted as a weighted average of precision and recall.

$$\text{F-Measure} = 2. \frac{\text{Precision. Recall}}{\text{Precision+Recall}} \tag{8}$$

— *Lift:* It tells about the predictive power of classifier arbitrary compared to a random selection by comparing the precision to the overall churn rate in the test set.

$$\text{Lift} = \frac{\text{Precision}}{P(P+N)} \tag{9}$$

— *Coverage:* the fraction of the number of cases satisfying both the condition and decision and number of cases satisfying decision only.

$$\text{Cov} = \frac{\text{Number of cases satisfying Condition and Decision}}{\text{Number of cases satisfying Decision}} \tag{10}$$

5   Evaluation Setup

In this section, an analytical environment is setup to perform the proposed technique using rough set explanation system (RSES) [34] and evaluated four different rules generation algorithms (i.e. exhaustive, genetic, covering, LEM2). These experiments carried out to fulfill the objectives of the proposed study and to address the following points also;

- **P1:** *Which features are more indicative for churn prediction in the telecom sector?*
- **P2:** *Which algorithm (Exhaustive, Genetic, Covering, and LEM2) is more appropriate for generating rules sets for rough set classification approach in telecommunication sector?*
- **P3:** *What is the predictive power of the proposed approach on churn prediction in the telecom sector?*
- **P4:** *Can derived rules help the decision makers in strategic decision making and planning process?*

5.1   Data Preparation and Feature Selection

Evaluating data mining approaches on publicly available dataset, has many benefits in term of comparability of results, ranking techniques, evaluating of existing methodologies with new one [40]. In this study, we have used publicly available dataset which can be obtained from URL [41].

Data preparation and feature selection are important steps in the knowledge discovery process, to identify those relevant variables or attributes from the large number of attributes in a dataset which are too relevant and that can reduce the computational cost [36]. The selection of most appropriate attributes from the dataset in hands, was carried out using feature ranking method titled as "Information Gain Attribute Evaluator", using an WEKA toolkit [42]. It evaluates the attributes worth through the information gain measurement procedure as per the class value. It's diverse the selection and ranking of attributes that significantly improves the computational efficiency and classification. After feature ranking, it includes most relevant and ranked attributes in the decision table. The Table 1 describes the selected attributes which are also addressed to P1.

Table 1. List of selected Top ranked attributes reflects the classification performance

Attributes	Description
Int'l Plan	Whether a customer subscribed international plan or not.
VMail Plan	Whether a customer subscribed Voice Mail plan or not.
Day Charges	A continuous variable that holds day time call charges.
Day Mins	No. of minutes that a customer has used in daytime.
CustSer Calls	Total No. of calls made a customer to customer service.
VMail Msg	Indicates number of voice mail messages
Int'l Calls	Total No. of calls that used as international calls.
Int'l Charges	A continuous variable that holds international call charges
Int'l Mins	No. of minutes that used during international calls.
Eve Charges	A continuous variable that holds evening time call charges.
Eve Mins	No. of minutes that a customer has used at evening time.
Churn?	The Class label whether a customer is churn or non-churn.

5.2   Preparation of Decision Table, Cut and Discretization

The preparation of decision table is an important stage of the proposed study of rough set theory based classification. The decision table which consists of objects, conditional attributes and decision attribute are organized in Table 2.

Table 2. Organization of attributes for decision table

Sets	Description
Objects	{ 3333 distinct objects}
Conditional Attributes	{Intl Phan, VMail_Plan, VMailMsg, Day_Mins, Day Charges, Eve_Mins, Eve_Charges, Intl_Mins, Intl_Calls, Intl_Charges, CustServ_Calls}
Decision Attribute	{Churn?}

Cut and discretization is the plausible approach to handle the large data by reducing the dataset horizontally. It is a common approach used in rough set where the variables which contains continuous values is partitioned into a finite number of intervals or groups. The cut and discretization process is carefully performed on the prepared decision table using RSES toolkits. It adds cuts in the subsequent loop one by one for a given attribute. It considers all the objects in decision table at every itera-tion and generate less number of cuts [34].

5.3   Training and Validation Sets

In data mining, validation is an extremely important step to ensure that the prediction model is not only remembering the instances that were given during the training process but it should also perform the same on unseen new instances. One way to overcome this problem is not to use the entire dataset for classifier's learning process. Some of the data are excluded from the training set as it begins the process to train the classifier. When the training process is completed, then excluded data can be used to validate the performance of the learned classifier on new data. This overall process of model evaluation is called cross validation. The performance of classifiers can be evaluated through several methods which have been discussed in literature [43].

In this study, the holdout cross validation method is used. We divided the data set into two sets: the training set and the test set. After multiple splitting attempts during the experiments, we concluded that the best performance is obtained if the split factor parameter is set to 0.7 using RSES toolkit that randomly splits the data into two dis-joint sub-tables. The division of the dataset into training and validation set are performed multiple times with irrespective of class labels and noted the average per-formance of the classifier to minimize the biases in the data.

5.4   Reduct and Decision Rules Sets Generation

The decision rules can be obtained from the training set by selecting either of the methods (Exhaustive, Genetic, Covering and LEM2). Where Exhaustive and Genetic algorithms are scanning the training set object-by-object and generate rules sets by matching the objects and attributes with reduct while Covering and LEM2 algorithms can induced rules sets without matching with reduct sets using RSES toolkit. In the proposed study, important decision rules are extracted from the training set through one-by-one these four different rules generations algorithms (Exhaustive, Genetic, Covering and LEM2). The decision rules set, specifies the rules in the form of "*if C then D*" where C is a condition and D refers to decision attribute. For example;

```
If Day_Mins=(108.8, 151.05) & Eve_Mins=(207.35, 227.15) &
    CustServ_Calls=(3.5, *)
Then  Churn=(True)
```

Based on these simple and easy interpretable rules, the decision makers can easily understand the flow of customer churn behavior and they can adopt more suitable strategic plan to retain their churn. All the generated decision rules induced from training set are summarized in Table 3.

Table 3. Statistics about rules induced using four methods

Description	Methods for Calculating Rules			
	Exhaustive	*Genetic*	*Covering*	*LEM2*
Total No. of Rules	4184	9468	369	625
# of rules induced that classifying customer as churn	1221	2674	122	160
# of rules induced that classifying customers as non-churn	2963	6715	247	465

6   Results and Discussion

This section reports the evaluation results and discussion on the performances of churn prediction classifiers which is observed during the experiments. The number of churns is much smaller as compared to non-churns customers in the selected dataset, which can provides tough time to churn prediction classifier during the learning process.

6.1   Evaluation of Classifiers' Performance

We have evaluated four different algorithms for rules generation through with rough set based classification approach using RSES toolkit. All these four methods are applied to the same telecom dataset. Table 4 reflects that genetic algorithm performed much better in term of obtaining 98% accuracy, 100% False churn and 98% true churn prediction along with coverage of all instances which is also reported to P2 because the results shows that it is a more appropriate algorithm out of target four. On the other hand, it is also investigated that genetic algorithm have shown more suitable predictive power which is addressed to P3.

Table 4. Evaluation of Four Rules Generation Mehtods through Rough Set Classification Approach

METHODS	TP	FP	FN	TN	COV	PRE	REC	ER	ACC	SPE	FM
Exhaustive	98	39	35	828	1	0.72	0.74	0.074	0.926	0.96	0.726
Genetic	**118**	**19**	**0**	**863**	**1**	**0.86**	**1.00**	**0.019**	**0.981**	**0.98**	**0.925**
Covering	37	41	37	525	0.64	0.47	0.50	0.122	0.878	0.93	0.487
LEM2	52	26	19	571	0.668	0.67	0.73	0.067	0.993	0.96	0.698

6.2   Comparison

Here we first address to P3: What is the predictive power of the proposed approach on churn prediction in the telecom sector? We have presented the sensitivity, specificity, precision, lift, misclassifications, accuracy and F-measure values in Table 4 and then compared the best performed method's prediction performance with other predictive models which are applied to the same dataset. Where A=Neural Network [2], B=Decision Tree [23], C=Neural Network [23], D=SVM [23], E= RBF kernel function of SVM [26], F= Linear kernel function of SVM [26], G= Polynomial kernel function of SVM [26], H=SIG kernel function of SVM [26] and I=Proposed Approach. By comparing the proposed approach with different Churn prediction techniques applied on similar data set, it is clear that the proposed approach performs very well as compared to the previously applied techniques.

Table 5. Predictive Performance of proposed & previous approaches applied to the same dataset

	A	B	C	D	E	F	G	H	I
Sensitivity	81.75	76.47	83.90	83.37	52.12	39.15	59.44	30.49	0.86
Specificity	94.70	79.49	83.50	84.04	94.70	95.32	96.29	93.12	1.00
Precision	66.27	80.60	83.40	84.20	81.90	79.05	80.95	71.43	1.00
Lift	0.0001	0.0002	0.0002	0.0002	0.0007	0.0005	0.0008	0.0004	0.00001
Type-1 Error	5.30	20.51	16.50	15.96	5.30	4.68	3.71	6.88	0.00
Type-2 Error	18.25	23.53	16.10	16.63	47.88	60.85	40.56	69.51	0.019
MisErr	6.76	22.10	16.30	16.30	14.37	22.14	11.44	29.47	0.019
Accuracy	93.2	77.9	83.7	83.7	85.6	77.9	88.6	70.5	0.981
F-Measure	73.2	78.5	83.7	83.8	63.7	52.4	68.5	42.7	0.925

6.3   Features' Analysis

As we have addressed P1 in sub-section 5.1, furthermore, we have examined the features' sensitivity to determine which features are more indicative for churn prediction in the telecom sector to address P1. Fig.1 shows the point of inflection of various variables such as CustServ_Call, Intl_Charges, Eve_Charges, Day_Charges, Day_Mins, Intl_Plan, Eve_Mins and above these points the churn rate decreases and the curve reflect the churn behavior while those features which are below the curve line e.g: Intl_Calls, VMail_Messages and VMail_Plan shows the churn rate constantly decreases. The curve is increasing at an increasing rate up to point of inflections and beyond these points churns behavior increasing at a decreasing rate. This implies that the churn rate is high in those features which are above the curve except Intl_Calls, VMail_Messages and VMail_Plan.

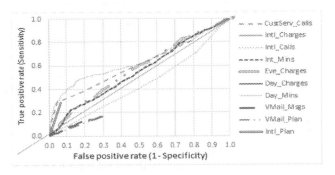

Fig. 1. Reflects sensitivity on y-axis and 1-specificity on x-axis

7   Conclusion

Customer churn is a crucial activity in rapidly growing and competitive telecommunication sector, due to the high cost of acquiring new customers. Churn prediction has emerged as an indispensable part of strategic decision making and planning process. This study is approaching to explore the powerful applications of rough set theory for churn prediction in telecommunication sector by constructing an improved predictive classifiers that can forecast churn prediction based on accumulated knowledge. To evaluate the results of the proposed technique, a benchmarking study is applied and as a result obtained not only more improved churn prediction accuracy, but also other evaluation measures as compared to other state-of-the-art techniques. In this study also investigated the performances of four different algorithms (Exhaustive, Genetic, Covering, and LEM2) of rules generation and extracted useful and easy interpretable decision rules. Based on these rules, the decision maker/manager can easily design more suitable strategic plan to retain the churn.

References

1. Hadden, J., Tiwari, A., Roy, R., Rute, D.: Computer assisted customer churn management: State-of-theart and future trends. IJCOR 10, 2902–2917 (2007)
2. Sharma, A., Kumar, P.: A Neural Network based Approach for Predicting Customer Churn in Cellular Network Services. IJCSA Application 27, 0975–8887 (2011)
3. Wouter, V., David, M., Christophe, M., Bart, B.: Building comprehensible customer churn prediction models with advance rule induction techniques. Expert Systems with Applications 38, 2354–2364 (2011)
4. Kirui, C., Hong, L., Wilson, C., Kirui, H.: Predicting Customer Churn in Mobile Telephony Industry Using Probabilistic Classifiers in Data Mining. IJCS 10 (2013)
5. Huang, B., Kechadi, M.T., Buckley, B.: Customer churn prediction in telecommunications. Expert Systems with Applications 39, 1414–1425 (2012)
6. Lina, C.S., Gwo-Hshiung, T., Yang Chieh, C.: Combined rough set theory and flow network graph to predict customer churn in credit card accounts. Expert System with Application 38, 8–15 (2011)
7. Yan, L., Wolniewicz, R.H., Dodier, R.: Predicting customer behavior in telecommunications. IEEE Intelligent Systems 2, 50–58 (2004)
8. Lazarov, V., Capota, M.: Churn Prediction, Business Analytics Course. TUM Computer Science (2007)
9. Den Poel, D.V., Lariviere, B.: Customer attrition analysis for financial services using proportional hazard models. European Journal of Operational Research, 196–217 (2004)
10. Chitra, K., Subashini, B.: Customer Retention in Banking Sector using Predictive Data Mining Technique. In: ICIT (2011)
11. Devi, P., Madhavi, S.: Prediction of Churn Behavior of Bank Customers Using Data Mining Tools. Business Intelligence Journal 5 (2012)
12. Tiwari, J., Hadden, A., Roy, R., Ruta, D.: Churn Prediction using Complaints Data. International Journal of Intelligent Technology 13, 158–163 (2006)
13. Lee, K.C., Chung, N.H., Kim, J.K.: A fuzzy cognitive map approach to integrating explicit knowledge and tacit knowledge: Emphasis on the churn analysis of credit card holders. Information Systems Review 11, 113–133 (2001)

14. Kawale, J., Aditya, Srivastava, J.: Churn prediction in MMORPGs: A social influence based approach. IEEE Computational Science and Engineering 4 (2009)
15. Suznjevic, M., Stupar, L., Matijasevic, M.: MMORPG player behavior model based on player action categories. In: 10th Workshop on NSSG. IEEE Press (2011)
16. Liou, J.J.H.: A novel decision rules approach for customer relationship management of the airline market. Journal of Expert Systems with Applications (2008)
17. Oentaryo, R.J., Lim, E.-P., Lo, D., Zhu, F., Prasetyo, P.K.: Collective Churn Prediction in Social Network. In: ASONAM. IEEE/ACM (2012)
18. Guo, L., Tan, E., Chen, S., Zhang, X., Zhao, Y.E.: Analyzing patterns of user content generation in online social networks. In: The 15th ACM SIGKDD, pp. 369–378 (2009)
19. Soeini, R.A., Keyvan, V.R.: Proceedings of Computer Science & Information Technology 30 (2012)
20. Burez, J., Van den Poel, D.: Handling class imbalance in customer churn prediction. Expert Systems with Applications 36, 4626–4636 (2009)
21. Ahn, J.-H., Han, S.P., Lee, Y.-S.: Customer churn analysis: Churn determinants and mediation effects of partial defection in the Korean mobile telecommunications service industry. Telecommunications Policy 30, 552–568 (2006)
22. Kim, M.K., Jeong, D.H.: The effects of customer satisfaction and switching barriers on customer loyalty in Korean mobile telecommunication services. Telecom Policy 28, 145–159 (2004)
23. Shaaban, E., Helmy, Y., Khedr, A., Nasr, M.: A Proposed Churn Prediction Model. IJERA 2, 693–697 (2012)
24. Qureshi, S.A., Rehman, A.S., Qamar, A.M., Kamal, A., Rehman, A.: Telecommunication Subscribers' Churn Prediction Model Using Machine Learning. IEEE (2013)
25. Kirui, C., Li, H., Cheruiyot, W., Kirui, H.: Predicting customer churn in mobile telephony industry using probabilistic classifiers in data mining. IJCSA 10, 1694–1814 (2013)
26. Innut, B., Churn, G.T.: Prediction in the telecommunications sector using support vector machines. Annals of Oradea University Fascicle of Mgt & Technological Engineering (2013)
27. Au, W.H., Chan, K.C., Yao, X.: A novel evolutionary data mining algorithm with applications to churn prediction. IEEE Trans. 7, 532–545 (2003)
28. Hossein, A., Mostafa, S., Tarokh, M.J.: The Application of Neuro-Fuzzy Classifier for Customer Churn Prediction. Procedia Information Technology & Computer Science 1, 1643–1648 (2012)
29. Farquad, M.A., Vadlamani, R., Raju, B.: Churn Prediction using comprehensible support vector machine: An Analytical CRM application. Elsevier Applied Soft Computing 19, 31–40 (2014)
30. Mozer, M., Wolniewicz, R., Grimes, D., Johnson, E., Kaushansky, H.: Predicting subscriber dissatisfaction and improving retention in the wireless telecommunications industry. IEEE Transactions on Neural Networks 11, 690–696 (2000)
31. Pawlak, Z.: Rough sets, rough relations and rough functions. Fundamenta informaticae 27, 103–108 (1996)
32. Pawlak, Z.: Rough sets. International Journal of Computer and Information Science 5, 341–356 (1982)
33. Zdzislaw, P.: Rough Set: theoretical aspects of reasoning about data. Kluwer Academic Publishers, Dordrecht (1991)
34. Bazan, J., Szczuka, M.S.: The rough set exploration system. In: Peters, J.F., Skowron, A. (eds.) Transactions on Rough Sets III. LNCS, vol. 3400, pp. 37–56. Springer, Heidelberg (2005)

35. Nguyen, H.S., Nguyen, S.H.: Analysis of stulong data by rough set exploration system (RSES). In: Proceedings of the ECML/PKDD Workshop (2003)
36. Bazan, J., Nguyen, H.S., Nguyen, S. H., Synak, P., Wróblewski, J.: Rough Set Algorithms in Classification Problem, pp. 49–88. Physica-Verlag, Heidelberg (2000)
37. Wroblewski, J.: Genetic algorithms in decomposition and classification problem. In: Skowron, A., Polkowski, L. (eds.) Rough Sets in Knowledge Discovery 1, pp. 471–487. Physica Verlag, Heidelberg (1998)
38. Grzymala-Busse, J.: A New Version of the Rule Induction System LERS. Fundamenta Informaticae 31, 27–39 (1997)
39. Burez, J.D., Van den Poel: Handling class imbalance in customer churn prediction. Expert Systems with Applications 36, 4626–4636 (2009)
40. Vandecruys, O., Martens, D., Baesens, B., Mues, C., De Backer, M., Haesen: Mining software repositories for comprehensible software fault prediction models. Journal of Systems and Software 81, 823–839 (2008)
41. Source of Dataset, http://www.sgi.com/tech/mlc/db/
42. Holmes, G., Donkin, A., Witten, I.H.: Weka: A machine learning workbench. In: Proceedings of the IEEE Intelligent Information Systems (1994)
43. John, H.: A Customer Profiling Methodology for Churn Prediction. P.hD thesis at Cranfield University (2008)

Analysis of Weather Information System in Statistical and Rough Set Point of View

Khu Phi Nguyen and Viet Long Huu Nguyen

University of Information Technology, Vietnam National University-HCMC,
Linh Trung, Thu Duc District, HCM City, Vietnam
khunp@uit.edu.vn, nguyenhuuvietlong@gmail.com

Abstract. Finding decision rules in weather prediction is an important issue besides using mathematical or statistical models and weather satellite data. It is dealt with this paper a method of analysing weather information system based on rough set reasoning and statistical method to support weather prediction. Firstly, some probability distributions of weather attributes will be verified and used for partition attribute value ranges using equal-probability percentiles. The solar terms in traditional East Asian point of view are used to determine types of decision weather attribute in the weather information system. Next, such an information system will be analyzed to produce a set of decision rules. After checking and verifying on collected weather dataset, it is shown that the method of analyzing is of significantly efficiency.

Keywords: Rough set, decision rules, prime implicant, probability distribution, statistical analysis.

1   Introduction

Almost people need weather forecasts at least once per day, weather significantly affects the health, safety, and socio-economic of everyone. Forecasting the weather requires continuously observing related data from the atmostphere and the underlying surfaces. Scientists have made considerable progress in developing computer programs to transform these observations into coherent analyses that serve as the foundation for weather prediction.

Traditionally, statistics have been used in the field of weather forecasting. A hots of studies have established statistical relationships between weather characteristic attributes in weather forecast, analysis of weather extremes and their impacts, [1]. Numerical methods are also used in weather forecasting. Numerical weather prediction [2] uses current weather conditions as input into mathematical models of the atmosphere to predict the weather. Numerical weather models [3]are initialized using observed data from surface weather observations or weather satellites.

Anyhow, analyses of collected weather data are worthy of attention, especially in information system point of view, [4-5].

© Springer International Publishing Switzerland 2015   97
D. Camacho et al. (eds.), *New Trends in Computational Collective Intelligence*,
Studies in Computational Intelligence 572, DOI: 10.1007/978-3-319-10774-5_9

This paper is dealt with analysing weather information systems using statistical method and rough set reasoning on collected weather data available at NCDC: http://www.ncdc.noaa.gov/cdo-web/datasets from the National Climatic Data Center. The paper is laid-out as follows: Section 2 introduces basic concepts and related works; Section 3 presents experimental computation results; Section 4 illustrates a case study related to the problem; and finally, a conclusion of the paper.

2 Basic Concepts

Recently, the rough set theory has been successfully applied to many real-life problems, [10, 14-17]. Rough set method has become a powerful tool for data mining, knowledge discovery in databases, especially in analyzing indiscernible, inconsistent, incomplete information systems. The main advantage of rough set theory is that it does not require additional information about data, [6].

2.1 Initial Definitions

Let Ω be a non-empty finite set of objects called universe, A be a finite set of attributes, then IS = (Ω, A) is an information system. An observation of such a system or an outcome of IS is represented by an information table or IT for short. The first column of IT illustrates objects, other columns attributes of each object.

Each $a \in A$ is a condition attribute with value set V_a of a, and corresponds with an information function $I_a: \Omega \rightarrow V_a$. Let Pv_a be a partition of V_a into a collection of disjoint subsets $V_{a,k}$, $k \in P_a = \{ 1, 2,\dots, p_a \}$, and $v_{a,k} \in V_{a,k}$ be a unique representative value of $V_{a,k}$. So, $I_a(\omega) \in V_{a,k}$ may be meant synonymously to $I_a(\omega) = v_{a,k}$.

For $B \subseteq A$, let $R_B \subseteq \Omega \times \Omega$ be a binary relation, defined as follows:

$$(\omega,\omega') \in R_B \Leftrightarrow \forall a \in B, \exists k \in P_a : I_a(\omega), I_a(\omega') \in V_{a,k} \tag{1}$$

In this case, R_B is an equivalence relation, called B-indiscernibility relation. An equivalence class of $\omega \in \Omega$ with respect to B is denoted by $[\omega]_B$. Set of all these equivalence classes is a partition of Ω, called the quotient space and notated by Ω/R_B.

Approximations to $X \subseteq \Omega$ are constructed on $B \subseteq A$ and R_B. B-lower and B-upper approximation of X, denoted by $B_L X$ and $B_U X$ respectively are defined in [7]:

$$B_L X = \{ \omega \mid [\omega]_B \subseteq X \}, \quad B_U X = \{ \omega \mid [\omega]_B \cap X \neq \varnothing \} \tag{2}$$

It is obtained $B_L X \subseteq X \subseteq B_U X$ and the difference $B_B X = B_U X - B_L X$ is called a B-boundary region of X. If the $B_B X$ is non-empty then X is rough, otherwise crisp.

A decision system is an information system (Ω, A) with a set of decision attributes D, notated by DS = $(\Omega, A\cup D)$. It is also obtained an equivalence relation R_D like R_B in (1) for $B\subseteq A$. The union of $B_L X$ for $X\in \Omega/R_D$ is the B-positive region with respect to D, denoted by $B_+(D)$.

In simple case D = {d} with $d\notin A$, let Pv_d be a partition of V_d into a collection of disjoint subsets $V_{d,k}\subset V_d$ $k\in P_d = \{1,2,\dots, p_d\}$, and let $v_{d,k}$ be representative value of $V_{d,k}$. Then the space Ω/R_d is a collection of disjoint subsets $\Omega_i = \{ \omega\in \Omega / I_d(\omega) = v_{d,i} \}$ with i = 1, 2,.. , p_d, called the ith decision class. The B-positive region with respect to d or $B_+(d)$ is the union of all $B_L(\Omega_i)$.

A decision system namely $(\Omega, A\cup\{d\})$ is consistent if $A_+(d) = \Omega$, otherwise inconsistent. The dependency between attributes is an important issue. Let B and C be attribute sets of A, it can be asserted that C depends on B if all values of attributes from C are uniquely determined by B, its degree of dependency is measured by:

$$\gamma_B(C) = |B_+(C)| / |\Omega| \tag{3}$$

Obviously, it is obtained $0 \leq \gamma_B(C) \leq 1$. If $\gamma_B(C) = 0$ then C does not depend on B. If $\gamma_B(C) = 1$, C depends totally on B and if $0 < \gamma_B(C) < 1$ C depends partially on B.

2.2   Reduction of Attributes and Decision Rules

A reduction of attributes, or briefly reduct, of decision system DS = $(\Omega, A\cup D)$ is a set B of attributes of A that preserves partitioning of Ω with A or $\Omega/R_A = \Omega/R_B$, hence $\gamma_A(D) = \gamma_B(D)$. An information system may have many reducts. A reduct with a minimal cardinality is called minimal reduct. Finding minimal reducts is a NP-hard problem, a subject of much research and many algorithms have taken interest in, e.g. QuickReduct, EBR or Entropy-Based Reduction, ACO algorithm [8, 9], etc.

For each $a\in B\subseteq A$, $\omega\in \Omega$ and $v\in V_a$, a proposition like $I_a(\omega) = v$, briefly $I_a = v$ or somewhere (a, v) takes a specific Boolean value. The notation $\phi := I_a = v$ is used to define a logic variable ϕ taking a truth of $I_a = v$, in other words, ϕ is true if there is an $\omega\in \Omega$ so that $I_a(\omega) = v$ or false otherwise. The meaning of ϕ denoted by $[\phi]$ is the set of all objects $\omega\in \Omega$ so that ϕ is true or $[\phi] = \{\omega\in \Omega \mid I_a(\omega) = v\}$.

The set $\Lambda(B)$ of logic variables and logic operations on B forms a set of logic expressions with respect to B, and called decision language from B. A decision rule in DS is a Boolean expression in the form: $\phi\rightarrow\psi$, read " if ϕ then ψ ", in which $\phi\in \Lambda(B)$ and $\psi\in \Lambda(D)$ are referred to as condition and decision of the rule, respectively.

The classification of a new object can be supported by matching its description to one of the decision rules. The support factor of the rule $\phi \rightarrow \psi$ can be determined by:

$$\text{Supp}(\phi, \psi) = | [\phi]\cap[\psi] | \tag{4}$$

Besides that, the certainty factor of $\phi \rightarrow \psi$ is also an important measure. This factor is usually called confidence coefficient, provided that $[\phi]\neq \varnothing$:

$$\mathrm{Cert}(\phi, \psi) = \mathrm{Supp}(\phi, \psi) \, / \, | \, [\phi] \, | \qquad (5)$$

It is obtained that $\mathrm{Cert}(\phi, \psi) = 1$ if and only if $\phi \rightarrow \psi$ is true, in this case, the rule is called a certain decision rule. If $0 < \mathrm{Cert}(\phi, \psi) < 1$ then $\phi \rightarrow \psi$ is uncertain.

Let $[\psi]$ be $\{\omega \in \Omega \mid I_d(\omega) = v_d \}$ with $\psi \in \Lambda(\{d\})$, $v_d \in V_d$. Assume that objects in $[\psi]$ and in the complement of $[\psi]$ are numbered with consecutive integral subscripts i and j, respectively. Decision matrix $M = (m_{ij})$ of a decision table with respect to $[\psi]$ is determined as a matrix whose entries are sets of attribute-value pairs, [11]:

$$m_{ij} = \{ \, (a, I_a(\omega_i)) \mid I_a(\omega_i) \neq I_a(\omega_j), \; a \in A \, \} \qquad (6)$$

The set m_{ij} consists of all attribute-value pairs in which values are not identical between ω_i and ω_j. The set of all minimal reducts of the collection of attribute-value pairs corresponding to row i, and the set of all minimal decision rules for that row can be obtained by forming the following Boolean expression, called a decision function: $B_i = \wedge_j (\vee_i m_{ij})$. Where $\wedge$ and $\vee$ are respectively generalized conjunction and disjunction operators. B_i is constructed out of row i of the decision matrix by taking conjunctions of disjunctions of pairs in each set m_{ij}.

The decision rules are obtained by turning each decision function into disjunctive normal form and using the absorption law of Boolean algebra to simplify it. The prime implicants of this simplified function correspond to the minimal decision rules. In fact, the number of prime implicants is often too large. Ranking the prime implicants via the support factor is a suitable way to identify the rule importance.

To find decision rules like if (basic weather elements) then (weather type), the following algorithm is applied:

```
Reasoning Algorithm
Input:   weather observation (basic weather elements)
Output: weather type
Begin
   Find in the Ranked Prime Implicants:
   if found return the corresponding weather type
      else continue with inconsistent rules
   Find in the Ranked Inconsistent Rules:
   if found return the corresponding weather type
            with its certainty value
      else return unknown weather type
End.
```

3   Analysis of Weather Dataset

Dataset on the weather of Ho Chi Minh City, Viet Nam have been collected from the data source available at http://www.ncdc.noaa.gov/cdo-web/datasets of the NCDC.

This dataset includes 4015 observations from 2003 to 2013, and 9 weather attributes: Average Temperature in °C, notated by T; Average Relative Humidity, H in %; Precipitation or Rainfall, R in mm; Average Wind Speed W in knot; Wind Direction, Wd; Station Level Pressure, P in millibar; Sea Level Pressure, Ps; Low Cloud, Cl in eighth; Total Cloud Cover, Ct. These attributes can be considered as random variables. Each of these variables conforms to a specified probability distribution.

In the traditional East Asian point of view, decision attribute of the weather information system is determined based on the weather characteristics of solar terms which match particular astronomical events or signifies some natural phenomena and are very important especially in agriculture.

Due to the solar terms are varied on region and time. It is necessary to have a way to identify when the spring is started or when the winter solstice is. Based on these times, decision attribute of weather information is assigned to weather type attribute which reflects the solar term value of a weather observation. Weather type currently focuses on 12 main solar term values, [13]: start of spring, notated by SSP; vernal equinox, VEQ; clear and bright, CAB; start of summer, SSU; summer solstice, SSS; minor and major heat, MMH; start of autumn, SAU; autumnal equinox, AEQ; frost descent, FDC; start of winter, SWI; winter solstice, WSS; minor and major cold, MMC.

3.1   Probability Distribution of Condition Attributes

In testing to identify an appropriate probability distribution of a given condition attribute, the value range of observed attribute data is partitioned into disjoint classes. The Goodness-of-Fit Chi-Square Test is used with a significant level at $\alpha = 5\%$, [12].

According to this test, the observed frequency O_i and expected frequency under the supposed probability distribution E_i of the ith class are evaluated. Then, the observed chi-square χ^2_{obs} - the sum of all $(O_i - E_i)^2/E_i$ is calculated. Let $\chi^2_{0.95,df}$ be the chi-square at 0.95-percentile and degree of freedom df = m-1; this number can be referred, for example, to Table AIV.4, pp. 766 in [12]. If $\chi^2_{obs} < \chi^2_{0.95,df}$ then the hypothesis on probability distribution of the attribute is accepted. The basic attributes T,P, H, W, and P are considered and results are illustrated in the following subsections.

Normal Probability Distributions. Using observed data, max-min or extremums of each attribute, its point estimation for population mean $\hat{\mu}$, and population standard deviation $\hat{\sigma}$ are calculated respectively:

- For T, extremums: 22.2-33.0 °C, $\hat{\mu} = 27.378$, $\hat{\sigma} = 1.495$, and $\chi^2_{obs} = 29.3928$, but $\chi^2_{0.95,22} = 33.924$; thus there is sufficient evidence to conclude that the probability distribution of T population is normal with mean $\hat{\mu}$ and variance $\hat{\sigma}^2$, $N(\hat{\mu}, \hat{\sigma}^2)$.

- For P, extremums: 1.0-34.0 millibar, $\hat{\mu} = 8.249$, $\hat{\sigma} = 1.997$, and $\chi^2_{obs} = 31.3683$, but $\chi^2_{0.05,26} = 38.885$; the distribution of P population is also normal $N(\hat{\mu}, \hat{\sigma}^2)$.

- For H, extremums: 45.6-100.4 mm, $\hat{\mu} = 78.289$, $\hat{\sigma} = 8.628$ and $\chi^2_{obs} = 28.5176$, while $\chi^2_{0.95,21} = 32.671$; the normality of distribution of H is accepted.

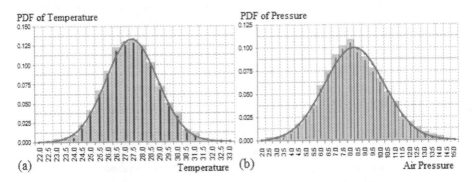

Fig. 1. Probability distribution of (a) T: Temperature and (b) P: Pressure

Gamma Probability Distributions. Calculations on observed data, corresponding characteristic numbers of W and R are estimated and given as follows:
- For W, extremums: 1.0 -53.4 km/hr, $\hat{\mu} = 4.485$, $\hat{\sigma} = 2.036$, and $\chi^2_{obs} = 21.9841$, while $\chi^2_{0.95,15} = 24.9960$; therefore, there is sufficient evidence to accept that the probability distribution of W values is Gamma $\Gamma(\hat{\alpha} = 4.850, \hat{\beta} = 0.925)$.
- For R, extremums: 0.0-900.0 mm, $\hat{\mu} = 5.843$, $\hat{\sigma} = 11.107$, and $\chi^2_{obs} = 20.7912$, $\chi^2_{0.95,14} = 23.6850$; its distribution is also Gamma $\Gamma(\hat{\alpha} = 0.2767, \hat{\beta} = 21.1132)$.

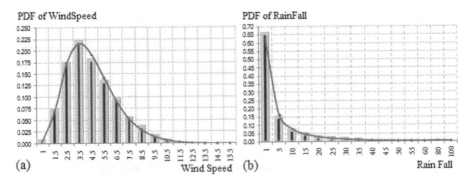

Fig. 2. Probability Distribution of (a) W: Wind speed and (b) R: Rainfall

3.2   Equal-Probability Discretization

Based on percentiles of the appropriate probability distribution of a given attribute variable and a given probability p of 5%, 10%, 20% a set of equal-probability cuts is determined, for example a set of equal-probability cuts obtained by using quartile. This set of cuts is used as an initial set to partition the value range of the attribute, [10]. Since then, it is proposed searches for an optimal set of cuts in a decision system such that a collection of all minimal set of cuts discerning objects in the DS. But this problem is NP-hard, hence Johnson strategy is applied to find an effective solution with an algorithm, named MD-heuristics [7].

The experimental result when running MD-heuristics in three levels of probability p and collected weather data from 2003 to 2012 is as follows:

- At p = 5%, there are 19 cuts for each attribute, 8 important attributes - P is reduced; and this case yields 8042 prime implicants.

- At p = 10%, there are 9 cuts per attribute, only 7 important attributes - P is also redundant; and it is obtained 4918 prime implicants.

- At p = 20%, there are 4 cuts per attribute, 8 important attributes - P is reduced; and this case gives 1723 prime implicants.

3.3   Decision Rules

Using the method for extracting rules illustrated in 2.2, the decison rules with support or certainty factors are found and presented in Table 1 and Table 2.

Table 1. Some of consistent decision rules with respect to p = 5%

Rule No.	If	then	Supp
1	$(76.1 \leq H < 78.4) \wedge (R < 0.4) \wedge (Wd = SW) \wedge (3.9 \leq Ws < 4.4)$	SSS	7
3	$(72.1 \leq H < 74.9) \wedge (Wd = SW) \wedge (5.4 \leq Ws < 6.8)$	SSU	6
4	$(28.4 \leq T < 28.7) \wedge (H < 65.1) \wedge (Wd = E)$	VEQ	6
6	$(26.6 \leq T < 27.1) \wedge (5.4 \leq Ws < 6.8) \wedge (CI = 3) \wedge$ $(1010.5 \leq P < 1010.9)$	MMC	5
22	$(6.8 \leq R < 9.2) \wedge (1010.5 \leq P < 1010.9)$	FDC	4
285	$(76.1 \leq H < 78.4) \wedge (CI = 7)$	VEQ	2
1310	$(65.1 \leq H < 68.3) \wedge (CI = 6)$	MMC	1

For inconsistent decision rules, it is received 34, 288, 508 rules in case p of 5%, 10%, 20% respectively. Table 2 shows some of inconsistent decision rules with p at 5% and certainty factors.

Table 2. Some of inconsistent decision rules with respect to p = 5%

Rule No.	If		then	Cert
1	$(28.7 \leq T < 29) \wedge (H < 65.1) \wedge (R < 0.4) \wedge (Wd = SE) \wedge$ $(3.9 \leq Ws < 4.4) \wedge (Ct = 2) \wedge (Cl = 2) \wedge (1008.2 \leq P < 1009)$		CAB	0.67
2	$(25.9 \leq T < 26.4) \wedge (85.4 \leq H < 88.6) \wedge (0.6 \leq R < 3.7) \wedge$ $(Wd = SW) \wedge (7.3 \leq Ws < 7.9) \wedge (Ct = 6) \wedge (Cl = 6) \wedge (P < 1005.3)$		SSS	0.67
4	$(28.7 \leq T < 29) \wedge (H < 65.1) \wedge (R < 0.4) \wedge (Wd = SE) \wedge$ $(3.9 \leq Ws < 4.4) \wedge (Ct = 2) \wedge (Cl = 2) \wedge (1009.5 \leq P < 1010.1)$		SAU	0.5
21	$(28.7 \leq T < 29) \wedge (72.1 \leq H < 74.9) \wedge (R < 0.4) \wedge (Wd = SW) \wedge$ $(7.3 \leq Ws < 7.9) \wedge (Ct = 3) \wedge (Cl = 3) \wedge (1005.3 \leq P < 1006)$		SAU	0.33

4   Verification

The received decision rules have been verified by using the dataset of 365 weather observations in 2013 in the same location of the NCDC website. Results of this verification are as follows and illustrated in Table 3.

Table 3. Verified prediction results using weather data in 2013 and p = 5%

Case No.	Observations	Weather type	Predicted weather type	Test
1	$T = 26.6 \wedge H = 72.8 \wedge R = 0 \wedge Wd = W \wedge$ $Ws = 3.6 \wedge Ct = 6 \wedge Cl = 2 \wedge P = 1012.7$	MMC	NOT FOUND	-
27	$T = 25.9 \wedge H = 66.1 \wedge R = 0 \wedge Wd = SE \wedge$ $Ws = 5.8 \wedge Ct = 6 \wedge Cl = 2 \wedge P = 1012.7$	MMC	MMC (rule 1310)	Good
29	$T = 26.2 \wedge H = 65.2 \wedge R = 0 \wedge Wd = SE \wedge$ $Ws = 5.7 \wedge Ct = 6 \wedge Cl = 2 \wedge P = 1012.7$	MMC	MMC (rule 1310)	Good
38	$T = 26.6 \wedge H = 76.6 \wedge R = 0 \wedge Wd = SE \wedge$ $Ws = 6.3 \wedge Ct = 6 \wedge Cl = 3 \wedge P = 1010.7$	SSP	MMC (rule 6)	Error
62	$T = 28.5 \wedge H = 61.4 \wedge R = 0 \wedge Wd = E \wedge$ $Ws = 6.5 \wedge Ct = 4.2 \wedge Cl = 4 \wedge P = 1009.6$	VEQ	VEQ (rule 4)	Good
68	$T = 27.3 \wedge H = 66 \wedge R = 0 \wedge Wd = SE \wedge$ $Ws = 8 \wedge Ct = 6 \wedge Cl = 4 \wedge P = 1012.4$	VEQ	MMC (rule 1310)	Error
90	$T = 26.8 \wedge H = 77.9 \wedge R = 0 \wedge Wd = NE \wedge$ $Ws = 8.6 \wedge Ct = 7 \wedge Cl = 7 \wedge P = 1008.1$	VEQ	VEQ (rule 285)	Good

Table 3. (*continued*)

125	T = 29.8 ∧ H = 72.9 ∧ R = 0 ∧ Wd = SW ∧ Ws = 6.7 ∧ Ct = 4 ∧ Cl = 4 ∧ P = 1005.6	SSU	SSU (rule 3)	Good
157	T = 26.3 ∧ H = 91 ∧ R = 1.8 ∧ Wd = SW ∧ Ws = 7.6 ∧ Ct = 6 ∧ Cl = 6 ∧ P = 1002	SSS	SSS (inconsistent rule 2)	Good
289	T = 27.1 ∧ H = 82.6 ∧ R = 7 ∧ Wd = SE ∧ Ws = 5 ∧ Ct = 5 ∧ Cl = 5 ∧ P = 1010.5	FDC	FDC (rule 22)	Good
...	...	...	...	...

- At p = 5%, there are 3 cases undefined by the obtained rules, but 322 of 362 observed cases agree with the decision rules, attain an accuracy of 89%.

- At p = 10%, 10 observed cases undefined, but 317 of 355 cases are agreed with the found decision rules, it is attained an accuracy of 89.3%.

- At p = 20%, this gives 16 cases undefined by the found rules, but 304 of 349 cases agreed with the extracted decision rules, and attain an accuracy of 87.1%.

5   Conclusion

In this paper, the rough set reasoning associated with statistical methods in analyzing weather information system have been considered. Identifying probability distributions of condition attributes plays an important role in equal-probability discretization and also in extraction decision rules for the system. Analyzing and verifying through case study shown that the proposed method is of significantly efficiency and worth doing on future research.

For future work, we intend to implement a rough ontology for weather information system in a given domain, and then, use such a rough ontology in characterizing and sharing weather knowledge as well as in weather prediction.

Acknowledgments. This work is funded by Vietnam National University-HCMC under grant number C2013-26-03. The authors are grateful to the National Climatic Data Center and Southern Regional Hydro-meteorological Center-Vietnam, in support of experimental data for this research.

References

1. Statistics of Weather and Climate Extremes (2014),
   http://www.isse.ucar.edu/extremevalues/extreme.html
2. Numerical Weather Prediction (2014),
   http://www.metoffice.gov.uk/services/industry/
   data/wholesale/model

3. Global Forecast System (2014),
   http://www.emc.ncep.noaa.gov/index.php?branch=modelinfo
4. Weather Information Systems, e.g. at (2014), http://www.weathermarket.com
5. Jayeoba, O.J., Amana, S.M., Ogbe, V.B.: Improving Weather Information Systems for Climate Change Assessment in Nigeria: The Role of AutomaticWeather Stations (AWSs), Publication of Nasarawa State University, PAT 9(1), 167–176 (2013), online copy available at http://www.patnsukjournal.net/currentissue ISSN 0794-5213
6. Hua, Q., Zhangb, L., Chenc, D., Pedryczd, W., Yua, D.: Gaussian kernel based fuzzy rough sets: Model, Uncertainty measures and Applications. Intl' Jour. of Approximate Reasoning (2010), doi:10.1016/j.ijar.2010.01.004
7. Pawlak, Z.: Andrzej Skowron: Rough sets: Some extensions. Information Sciences (177), 28–40 (2007)
8. Jensen, R.: Combining rough and fuzzy sets for feature selection. Dr. Thesis, pp. 112–120. School of Informatics, University of Edinburgh (2005)
9. Chen, Y., Miao, D., Wang, R.: A rough set approach to feature selection based on ant colony optimization. Pattern Recognition Letters 31, 226–233 (2010), doi:10.1016/j.patrec.2009.10.013
10. Nguyen, K.P., Bui, S.T., Tu, H.T.: Revenue Evaluation Based on Rough Set Reasoning. In: Sobecki, J., Boonjing, V., Chittayasothorn, S. (eds.) Advanced Approaches to Intelligent Information and Database Systems. SCI, vol. 551, pp. 159–169. Springer, Heidelberg (2014)
11. Nguyen, K.P., Tu, H.T.: Data mining based on Rough set theory. A chapter in the book: Knowledge Discovery in Databases. Academy Publisher Inc., USA (2013)
12. Kandethody, M., Ramachandran, C.P.: Tsokos: Mathematical statistics with applications, Elsevier Academic Press, San Diego, California 92101-4495, USA (2009)
13. Solar term (2014), http://en.wikipedia.org/wiki/Solar_term
14. Lee, M.-C.: An Enterprise Financial Evaluation Model Based on Rough Set theory with Information Entropy. Intl' J. of Digital Content Technology and its Applications 3(1), 16–22 (2009)
15. Chen, Y., Wang, S., Chan, C.-C.: Application of Rough Sets to Patient Satisfaction Analysis. In: Proc. of the 11th Intl' DSI and the 16th APDSI Joint Meeting, Taipei, Taiwan (2011)
16. Ge, Y., Cao, F., Du, Y., Lakhan, V., Wang, Y., Li, D.: Application of Rough Set-Based Analysis to Extract Spatial Relationship Indicator Rules: An example of Land Use in Pearl River Delta. J. Geogr. Science 21(1), 101–117 (2011)
17. Kozae, A.M., El-Sheikh, S.A., Aly, E.H., Hosny, M.: Rough sets and its applications in a computer network. Annals of Fuzzy Mathematics and Informatics (2013), http://www.afmi.or.kr, ISSN 2093-9310

Exclusive Channel Allocation Methods Based on Four-Color Theorem in Clustering Sensor Networks

Mary Wu[1], SungYong Ha[2], Turki Abdullah[3], and ChongGun Kim[3,*]

[1] Dept. Of Computer Culture,
Yongnam Theological University and Seminary, Korea
[2] Dept. Of Computer Information,
Kyungbuk College, Korea
[3] Dept. Of Computer Engineering,
Yeungnam University, Korea
mrwu@ynu.ac.kr, hsy@kbc.ac.kr,
Price.turki.1988@gmail.com, cgkim@yu.ac.kr

Abstract. The effective use of energy efficiency is a big challenge in sensor networks, since sensor nodes have limited power and are deployed in inaccessible regions. Clustering on the sensor networks reduces the volume of inter-node communications and raises energy efficiency by transmitting the collected data from members by a cluster head to a sink node. Nevertheless, due to radio frequency characteristics in wireless communications, interference and collision can occur between neighboring clusters, leading to increased energy consumption due to the resulting re-transmission. The occurring of interference and collision channels among clusters can be resolved by assigning non-overlapping channels among neighbor clusters. In this paper, we propose an exclusive channel allocation method based on four-color theorem. This method assigns exclusive channels among neighbor clusters in various clustered topologies and makes nodes transfer data in collision free environments among clusters. The experimental results show successful assigning of exclusive channel among neighbors using the 4 number of channels in various cluster topologies.

Keywords: Cluster sensor networks, interference among clusters, exclusive channel allocation, four-color theorem, neighbor matrix.

1 Introduction

Wireless Sensor Networks collect data on the surrounding environment and can be applied to a variety of purposes such as intrusion detection in military areas, area security, and environmental monitoring of temperature and humidity. Sensor nodes become aware of the resulting symptoms and transmit the measured data to a base station, which in turn analyzes the data. A limitation arises due to the limited resources of sensor nodes on the wireless sensor networks. Many studies on the efficient use of energy have been conducted to overcome this problem[1-7].

---

* Corresponding author.

© Springer International Publishing Switzerland 2015
D. Camacho et al. (eds.), *New Trends in Computational Collective Intelligence*,
Studies in Computational Intelligence 572, DOI: 10.1007/978-3-319-10774-5_10

Typically, neighboring sensor nodes collect similar information, leading to large energy wastage because of duplicated transmission of similar information. Consequently, many cluster methods on sensor networks have been studied. Clustering, in which the sensor network is divided into non-overlapping groups of nodes, is an effective method for achieving high levels of energy efficiency and scalability. In clustering, each node belongs to a local cluster and the cluster head integrates the data collected from members of the cluster, and then transmits it to a sink node. This prevents duplicated transmission of similar information and gives low-power networking in the sensor networks[5-7].

Transmission synchronization among clusters can achieve through the assignment of different resources among neighbor clusters[8-9]. In the Low-Energy Adaptive Clustering Hierarchy (LEACH)[2], a typical clustering protocol for sensor networks, cluster heads make the time division multiple access (TDMA) schedule for their cluster members and allocate it to each member node. Because the TDMA schedule of each cluster is built independently, the data transmission of nodes located near the cluster boundary can cause collisions with the transmission of a neighboring cluster.

To minimize this type of interference, each cluster communicates using different code division multiple access (CDMA) codes in LEACH[2]. This is not a suitable choice for implementing CDMA code in sensor nodes at low cost.

TDMA-based Avoiding Collision (TAC)[10] solves this problem in the view of TDMA. Clusters exchange control messages for the allocation of non-overlapping group IDs among adjacent clusters based on a hexagonal cluster model. The Initial cluster broadcasts a group allocation message including its own group ID and the received clusters allocate the group ID which isn't same group ID compared with that of the received message. This process is done gradually and repeatedly until all clusters have non-overlapping group IDs in the networks. This method requires the exchange of many messages and preprocess of time synchronization.

In our previous study[11-12], we proposed a channel reuse procedure which assigns dynamically channels that do not overlap among adjacent clusters in cluster sensor networks using the matrices, such as an adjacency matrix, a topology matrix and a resource allocation matrix, based on a hexagonal cluster model. The complicate calculations which are required in this method are made by a gateway or a server with non-limited amount of memory, high power and high processing capability. Therefore, the method is well suited, when considering the efficient use of the energy of sensor nodes, because it doesn't require the exchange of many messages among sensor nodes. But this method uses the hexagonal model, it doesn't completely solve the collisions among neighboring clusters in various cluster topologies.

In this study, therefore, we propose an exclusive channel allocation method for polygonal clustered networks. It is based on the four-color theorem, which states any separation of a plane into contiguous regions and no two adjacent regions have the same color. Two regions are called adjacent if they share a common boundary that is not a corner, where corners are the points shared by three or more regions[13-17]. Our method assigns successfully exclusive channels among neighbors using only 4 channels in various cluster topologies.

2   Relates Works

TAC protocol[10] solves the problems of the interference and the collision through the allocation of the different time periods among neighbor clusters. In the study, a cluster topology uses a hexagonal model and each cluster is assigned using the different group ID among neighbor cluster from '0' to '2'.

This method assigns channel '0', '1' to an initial cluster, the neighbor cluster of it, respectively and then, assigns the different channel number with the two within '0' - '2' to the neighbor cluster of both. This process is continued repeatedly until channels are assigned to the entire clusters. A large amount of messages are exchanged and an amount of energy is consumed.

[11-12] propose the calculation methods for the allocation of non-overlap channels among neighbor clusters based on a hexagonal model.

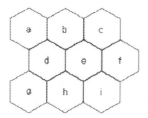

Fig. 1. Cluster topology based on a hexagonal model

$$T = \begin{bmatrix} t_{11} & t_{12} & t_{13} & t_{14} & t_{15} & t_{16} \\ t_{21} & t_{22} & t_{23} & t_{24} & t_{25} & t_{26} \\ t_{31} & t_{32} & t_{33} & t_{34} & t_{35} & t_{36} \end{bmatrix} = \begin{bmatrix} a & 0 & b & 0 & d & 0 \\ 0 & d & 0 & e & 0 & f \\ g & 0 & h & 0 & i & 0 \end{bmatrix}. \quad (1)$$

$$RA = \begin{bmatrix} 1 & \infty & 0 & \infty & 2 & \infty \\ \infty & 2 & \infty & 1 & \infty & 0 \\ 1 & \infty & 0 & \infty & 2 & \infty \end{bmatrix},$$

$$where \quad ra_{ij} = \begin{cases} (3i + j) \ \% \ 3, & for \ t_{ij} \in T, \ if \ t_{ij} \neq 0, \\ \infty, & otherwise . \end{cases} \quad (2)$$

The method uses the adjacent matrix based on the adjacent relation of the clusters, T(cluster topology matrix), RA(the resource allocation matrix). The (1) shows T for the cluster topology of the fig. 1 and the (2) shows RA for it. The calculation of the matrices is made by the server or gateway which is not constrained by performance factors, such as the amount of memory, power, processing capacity. But, it is based on the hexagonal clustered model, overlapping channels among neighbor clusters may be not assigned in various clustered topologies.

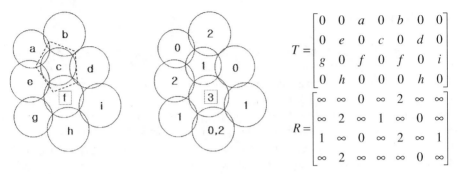

Fig. 2. A non-hexagonal cluster topology and matrices

Fig. 2 shows that the number of neighbors of cluster 'c' is 5. In the non-hexagonal model, cluster 'h', 'f' are presented twice in the topology matrix. According to the rule of resource allocation matrix, cluster 'f', 'h' are allocated by the channel number '0' or '2' by RA (2). Because the channels '0', '2' are already allocated to the neighbors of cluster 'f', an additional channel '3' is reallocated to the cluster 'f'. In various and complicate cluster topologies, many clusters can be presented more than twice in the topology matrix T and the ambiguity and the complexity for the allocation of non-overlapping channels increases. More clear and more simple method is needed.

3   Exclusive Channel Allocation Methods

3.1   Channel Allocation Algorithm

Based on the four-color theorem, the exclusive channel among adjacent clusters can be assigned by using 4 channels in some form of topology. Fig. 3(b) represents the results of the exclusive channel allocation using the four numbers 1,2,3,4 in the topology of fig. 3(a).

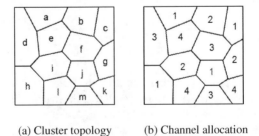

(a) Cluster topology   (b) Channel allocation

Fig. 3. Cluster topology and the result of exclusive channel allocation

For exclusive channel allocation among neighbors in the clustered sensor network, we use the adjacency matrix A which presents the adjacency relation among clusters and the exclusive channel matrix EC which presents the process of exclusive channel allocation based on the A.

(3) shows the A matrix of the cluster topology of fig. 3(a). In the A, when two clusters are adjacent, the elements are represented by 'n' and two clusters are not adjacent the elements are represented by '0'[18].

$$
A = \begin{array}{c}
\\ a \\ b \\ c \\ d \\ e \\ f \\ g \\ h \\ i \\ j \\ k \\ l \\ m
\end{array}
\begin{array}{ccccccccccccc}
a & b & c & d & e & f & g & h & i & j & k & l & m \\
\infty & n & 0 & n & n & 0 & 0 & 0 & 0 & 0 & 0 & 0 & 0 \\
n & \infty & n & 0 & n & n & 0 & 0 & 0 & 0 & 0 & 0 & 0 \\
0 & n & \infty & 0 & 0 & n & n & 0 & 0 & 0 & 0 & 0 & 0 \\
n & 0 & 0 & \infty & n & 0 & 0 & n & n & 0 & 0 & 0 & 0 \\
n & n & 0 & n & \infty & n & 0 & 0 & n & 0 & 0 & 0 & 0 \\
0 & n & n & 0 & n & \infty & n & 0 & n & n & 0 & 0 & 0 \\
0 & 0 & n & 0 & 0 & n & \infty & 0 & n & 0 & n & 0 & 0 \\
0 & 0 & 0 & n & 0 & 0 & 0 & \infty & n & 0 & 0 & n & 0 \\
0 & 0 & 0 & n & n & n & 0 & n & \infty & n & 0 & n & 0 \\
0 & 0 & 0 & 0 & 0 & n & n & 0 & n & \infty & n & n & n \\
0 & 0 & 0 & 0 & 0 & 0 & n & 0 & 0 & n & \infty & 0 & n \\
0 & 0 & 0 & 0 & 0 & 0 & 0 & n & n & n & 0 & \infty & n \\
0 & 0 & 0 & 0 & 0 & 0 & 0 & 0 & 0 & n & n & n & \infty
\end{array}
\tag{3}
$$

The following shows the exclusive channel allocation algorithm. The notations for the algorithm are defined as follows;

AC = { x | x is a cluster in WSN}
NC(k) = { x | x is the neighbor of k cluster}

AC is the set of all the cluster in WSN and NC(k) is the set of the neighbor clusters of cluster k.

The cluster with the minimum number of neighboring nodes is selected as an initial FR cluster. In the adjacency matrix A, 'a', 'c', 'h', 'k', and 'm' rows have three neighbors which are the minimum number of neighbors. One of the 5 clusters, an arbitrary cluster 'a', is selected as the initial FR and the channel number '1' and an allocation sequence number '1' are allocated to it.

The neighboring clusters of the initial FR 'a' are 'b', 'd', and 'e' in A. One of them is selected as initial SR and the channel number '2' and the allocation sequence number '2' are allocated to it.

In the partial EC matrix of fig. 4(b), the channel number '1', '2' are allocated to the cluster 'a', 'b' and the columns 'a', 'b' are set to '1','2', respectively.

An exclusive channel is allocated to the neighbor of the FR and the SR. In the FR row and SR row of N matrix, if there are columns which have the same 'n' elements, select one of the columns.

The column 'e' has the 'n' elements in the FR 'a' row and the SR 'b' row of EC matrix in the fig. 4(c). The column 'e' is selected as a channel allocation cluster. The exclusive channel number and the allocation sequence number are assigned to it.

---

1.   Elect one of the clusters which have the minimum neighbor as an initial FR(First Reference) and the channel number '1' and an allocation sequence number '1' are allocated to it.
2.   Elect one of the clusters which are the neighbors of the FR as an initial SR(Second Reference) and the channel number '2' and the allocation sequence number '2' are allocated to it.
3.   Neighbor cluster channel allocation based on FR and SR.
   If there are columns which have the 'n' elements in the FR and SR rows of N matrix,

   ① Selects one of the columns.

   ② Assigns the channel number to the selected column by using the channel allocation rule and assigns the allocation sequence number which is increased by a value of 1 to it.

   ③ The cluster which is assigned a channel becomes the new SR.

   ④ Go 3.

   Else if there are columns which have the same channel number element in the FR and SR rows of N matrix,
      If channels aren't assigned to the all element of NC(FR),

      ① The column becomes SR.

      ② Go 3.

   Else if channels are assigned to the all element of NC(FR),
      If channels aren't assigned to the all element of AC,

         ① A new FR is selected based on the ascending order of the allocation sequence number.

         ② One assigned neighbor of the new FR is selected as the new SR.

         ③ Go 3.

      Else

         Finish.
   Else if channels aren't assigned to the all element of NC(FR),

         ① One of NC(FR) which aren't assigned channels is selected, and another channel with that of FR is allocated to it

         ② The column becomes SR.

         ③ Go 3.

We made the channel allocation formula (4) to assign non-overlapping channels among adjacent clusters. The channel number is used in the range of from 1 to 4.

$$(3i + 2j) \% 4 + 1, \tag{4}$$

where i, j are the channel numbers which are allocated to the two reference clusters.

Table 1. Channel allocation numbers based on the channel number of FR and SR

i	j	The results of the channel allocation formula	The second channel number
1	2	4	3
2	3	1	4
3	4	2	1
4	1	3	2
1	3	2	4
2	4	3	1

Table 1 shows the results of the channel allocation formula and the second channel number which can be replaced when there already exists the same channel number to the neighbors. The channel number ‘4’ is allocated to the cluster ‘e’ by the channel allocation rule of table 1 and the results shows in the fig. 4(d). The allocation sequence number ‘3’ is allocated to it.

The cluster ‘e’ which is assigned a channel becomes the new SR. The exclusive channels allocation based on the FR and the SR is repeated.

In the fig. 4(e), the column ‘d’ has the ‘n’ element in the rows of the FR ‘a’ and the SR ‘e’.

The column ‘d’ is selected as the channel allocation cluster. The exclusive channel number ‘3’ is allocated to the cluster ‘d’ by the channel allocation rule in the fig. 4(f). The allocation sequence number ‘4’ is allocated to it.

The cluster ‘d’ becomes the new SR for the FR. If exclusive channels are assigned to all the neighbors of FR, ‘a’, the new FR must be selected based on the allocation sequence number. The cluster ‘b’ of which the allocation sequence number is ‘2’ is selected as the next FR.

$$A = \begin{array}{c} \\ a \\ b \\ d \\ e \end{array} \begin{array}{cccc} a & b & d & e \\ \infty & n & n & n \\ n & \infty & 0 & n \\ n & 0 & \infty & n \\ n & n & n & \infty \end{array}.$$

(a) A partial A matrix

$$EC = \begin{array}{c} \\ a\,\boxed{1} \\ b\,\boxed{2} \\ d\ n \\ e\ n \end{array} \begin{array}{cccc} a & b & d & e \\ \infty & 2 & n & n \\ 1 & \infty & 0 & n \\ 1 & 0 & \infty & n \\ 1 & 2 & n & \infty \end{array}.$$

(b) A partial EC matrix

$$EC = \begin{array}{c} \\ a\ 1 \\ b\ 2 \\ d\ n \\ e\ n \end{array} \begin{array}{cccc} a & b & d & e \\ \infty & 2 & n & \boxed{n} \\ 1 & \infty & 0 & \boxed{n} \\ 1 & 0 & \infty & n \\ 1 & 2 & n & \infty \end{array}.$$

(c) A partial EC matrix

$$EC = \begin{array}{c} \\ a\ 1 \\ b\ 2 \\ d\ n \\ e\,\boxed{4} \end{array} \begin{array}{cccc} a & b & d & e \\ \infty & 2 & n & 4 \\ 1 & \infty & 0 & 4 \\ 1 & 0 & \infty & 4 \\ 1 & 2 & n & \infty \end{array}.$$

(d) A partial EC matrix

$$EC = \begin{array}{c} \\ a\ 1 \\ b\ 2 \\ d\ n \\ e\ 4 \end{array} \begin{array}{cccc} a & b & d & e \\ \infty & 2 & \boxed{n} & 4 \\ 1 & \infty & 0 & 4 \\ 1 & 0 & \infty & 4 \\ 1 & 2 & \boxed{n} & \infty \end{array}.$$

(e) A partial EC matrix

$$EC = \begin{array}{c} \\ a\ 1 \\ b\ 2 \\ d\,\boxed{3} \\ e\ 4 \end{array} \begin{array}{cccc} a & b & d & e \\ \infty & 2 & 3 & 4 \\ 1 & \infty & 0 & 4 \\ 1 & 0 & \infty & 4 \\ 1 & 2 & 3 & \infty \end{array}.$$

(f) A partial EC matrix

Fig. 4. Channel allocation process using an exclusive channel algorithm

The process of exclusive channel allocations is repeated, until exclusive channels are allocated to all the clusters in WSN.

4   Experiments

The experiments for the performance of the exclusive channel allocation are performed. Fig. 5(b-d) shows that examples of channel allocations using the TAC algorithm, T and RA matrices, the four-color method based on the cluster topology fig.5(a).

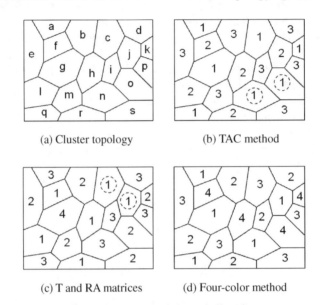

(a) Cluster topology   (b) TAC method

(c) T and RA matrices   (d) Four-color method

Fig. 5. The results of channel allocation

Fig. 6. The results of four color channel allocation algorithm

Fig. 5(b) shows the results of TAC channel allocation. Cluster 'n', and cluster 'o' have the same channel number as '1'. Fig. 5(c) shows the results of T matrix and RA matrix channel allocation. Cluster 'c', and cluster 'j' have the same channel number as '1'. The result of the fig. 5(d) shows that every channel is exclusive among neighbors.

Fig. 6 shows the results of our proposed four color channel allocation algorithm. The proposed algorithm successfully allocates non-overlapping channels by using four channels in complex topologies.

5   Conclusions

The inter-cluster synchronization of channels in the cluster-based sensor networks is important. In this paper, the channel allocation method based on the four-color theorem is proposed to provide exclusive channels among neighbor clusters. The execution of the proposed algorithm shows the result of successful exclusive channel allocation among adjacent clusters in various cluster. In future research, the practical protocol and procedures that can be applied in real clustered wireless network systems must be studied.

References

1. Akyildiz, I.F., Su, W., Sankarasubramaniam, Y., Cayirci, E.: Wireless sensor networks: a survey. Computer Networks 38 (2002)
2. Heinzelman, W.R., Chandrakasan, A., Balakrishnan, H.: Energy-Efficient Communication Protocol for Wireless Microsensor Networks. In: Proceedings of the Hwaii International Conference on System Science, pp. 1–10 (January 2000)
3. Ye, W., Heidemann, J., Estrin, D.: An Energy-Efficient MAC Protocol for Wireless Sensor Networks. In: Proceedings of the 21st IEEE INFOCOM, vol. 3, pp. 1567–1576 (June 2002)
4. Bhanumathi, V., Dhanasekaran, R.: Path Discovery And Selection For Energy Efficient Routing With Transmit Power Control In MANET. Malaysian Journal of Computer Science 26(2) (2013)
5. Ying, L., Haibin, Y.: Energy Adaptive Cluster-Head Selection for Wireless Sensor Networks. In: Proceedings of the 6th International Conference on Parallel and Distributed Computing Applications and Technologies, pp. 634–638 (December 2005)
6. J.: 3DE: Selective Cluster Head Selection scheme for Energy Efficiency in Wireless Sensor Networks. In: The 2nd ACM International Conference on Pervasive Technologies Related to Assistive Environments (2009)
7. Hong, T.-P., Wu, C.-H.: An Improved Weighted Clustering Algorithm for Determination of Application Nodes in Heterogeneous Sensor Networks. Journal of Information Hiding and Multimedia Signal Processing 2(2), 173–184 (2011)
8. Marsh, G.W., Kahn, J.M.: Channel Reuse Strategies for Indoor Infrared Wireless Communications. IEEE Transactions on Communications 45(10) (October 1997)
9. Wang, X., Berger, T.: Spatial channel reuse in wireless sensor networks. Journal of Wireless Networks 14(2) (March 2008)

10. Leem, I.T., Wu, M., Kim, C.: A MAC scheme for avoiding inter-cluster collisions in wireless sensor networks. In: 2010 The 12th International Conference on Advanced Communication Technology(ICACT), February 7-10, pp. 284–288 (2010)
11. Wu, M., Leem, I., Jung, J.J., Kim, C.: A Resource Reuse Method in Cluster Sensor Networks in Ad Hoc Networks. In: Pan, J.-S., Chen, S.-M., Nguyen, N.T. (eds.) ACIIDS 2012, Part II. LNCS, vol. 7197, pp. 40–50. Springer, Heidelberg (2012)
12. Wu, M., Ahn, B., Kim, C.: A Channel Reuse Procedure in Clustering Sensor Networks. Applied Mechanics and Materials 284-287, 1981–1985
13. Robertson, N.: The Four-Colour Theorem. Journal of Combinatorial Theory, Series B 70, 2–44 (1997)
14. Bar-Natan, D.: Lie Algebra and The Four Color Theorem. Combimatorica 17(1), 43–52 (1997)
15. Thomas, R.: An Update On The Four-Color Theorem. Notices Amer. Math. Soc. 45, 848–859 (1998)
16. Eliahow, S.: Signed Diagonal Flips and the Four Color Theorem. Europ. J. Combinatorics 20, 641–647 (1999)
17. Hodneland, E., Tai, X.-C., Gerdes, H.-H.: Four-Color Theorem and Level Set Methods forWatershed Segmentation. Inter. J. Comput. Vis. 82, 264–283 (2009)
18. Wu, M., Kim, C.: A cost matrix agent for shortest path routing in ad hoc networks. Journal of Network and Computer Applications 33, 646–652 (2010)

3D Human Face Recognition Using Sift Descriptors of Face's Feature Regions

Nguyen Hong Quy[1], Nguyen Hoang Quoc[1], Nguyen Tran Lan Anh[1],
Hyung-Jeong Yang[2], and Pham The Bao[1]

[1] Faculty of Math and Computer Science, University of Science, Ho Chi Minh City, Vietnam
[2] Department of Computer Engineering, Chonnam National University, South Korea
nhquy@hotmail.com, {nhquoc,hyungjeong}@gmail.com,
{ngtlanh,ptbao}@hcmus.edu.vn

Abstract. Many researches in 3D face recognition problem have been studied because of adverse effects of human's age, emotions, and environmental conditions on 2D models. In this paper, we propose a novel method for recognizing 3D faces. First, a 3D human face is normalized and determined regions of interest (ROI). Second, SIFT algorithm is applied to these ROIs for detecting invariant feature points. Finally, this descriptor, extracted from a training image, will be stored and later used to identify the face in a test image. For performing reliable recognition, we also adjust parameters of SIFT algorithm to fit own characteristics of the template database. In our experiments, the proposed method produces promising performance up to 84.6% of accuracy when using 3D Notre Dame biometric data-TEC.

Keywords: 3D face recognition, SIFT descriptors, range images.

1 Introduction

Nowadays, many fields such as finance, banking, stock market, etc. require high level of security. The need of fast and precise human identification in business transactions has become urgent. A lot of biometric technologies (e.g. fingerprint, iris, face) are exploited due to their high reliable characteristics. Although human face contains less invariant features than others, it still gains much potential and suitable low-price security applications. Currently, face recognition systems only focus on 2D images taken by normal digital cameras. Despite of achieving very good results, it cannot satisfy researchers because these 2D images may flatten special features appearing in the face as well as there exists restrictions caused by objective effects of light, noise, facial emotions, etc. So its important depth information may be lost. For this reason, 3D human face recognition algorithms have been studied more and more since not only faces captured by 3D models contain a lot of information but also they are not affected by negative effects.

As clarified in surveys of 3D face recognition methods, their main researches can be divided into two categories [1]: one processes only pure 3D data and the other is a

© Springer International Publishing Switzerland 2015
D. Camacho et al. (eds.), *New Trends in Computational Collective Intelligence*,
Studies in Computational Intelligence 572, DOI: 10.1007/978-3-319-10774-5_11

combination of 2D and 3D data. Approaches belonging to the second type were promoted quite late in 2000. Most of them try to take results obtained from both 2D and 3D models for producing better conclusions. Wang et al. in [2] presented an idea of this combination by describing feature points using Gabor filter responses in a 2D gray-level image and point signature in a 3D domain. Later, an approach proposed by Gang Pan et al. [3] was to automatically extract ROI of facial surface by considering bilateral symmetry plane and localization of nose tip. In [4], [5], [6], [7] Principal Component Analysis (PCA) algorithm and its modified versions were applied to get principal components of range images as feature points. After that, various distance measurements were used for 3D face recognition or classification. J. Cook et al. [8] combined Log-Gabor Templates for variation expression of depth and texture data with Mahalanobis Cosine metric as the distance measure for solving the problem. Suggested in [9] by Pamplona Segundo et al., a scale-invariant face detection approach were proposed by using boosted cascade classifiers in range images as input for real-time 3D face recognition system. Jahanbin S et al. [10] motivated a novel multimodal framework based on local attributes such as 2-D and 3-D Gabor coefficients and the anthropometric distances. Finally three parallel face recognizers were considered to form a recognition system at matching score level. However processing very large volume of 3D information is still the most challenge of 3D recognition models. Important facial characteristics can be skipped for reducing processing time during this stage. For adapting to real-time systems, a balance of accuracy and quickness in 3D face recognition algorithms is needed. Derived from above issues, developing a simple and high reliable 3D recognition model is the key point of our proposed method.

In this paper, a novel face recognition method by applying a 2D processing technique into pure 3D face images is represented. In Section 2, we describe the proposed method in details. Section 3 gives experiments on 3D Notre Dame biometric data-TEC database. Finally, conclusions are drawn from our work in Section 4.

2   Methodology

2.1   General System

To describe points of a face in our 3D model, data is input in the form of three matrices and position of the top of its nose is always given. An example of our input data is shown in Fig. 1 after mapping three data matrices to a 2D face image.

Fig. 1. An input face after mapping from 3D to 2D space including its depth value

2.2   Face Normalization

When sampling data, it is very difficult and inconvenient to capture well human faces as our standard criteria. We might accept samples that have slight deviation in both three dimensions. Although this flexibility does not affect SIFT descriptors [15], the input should be rotated into its front face after this step for an easy ROI extraction.

Horizontal Rotation. For rotating a face image horizontally, we try to balance the deviation of two tops of nostrils illustrated as red points in Fig. 2. Since position of the top of the nose is given, it is quite easy to determine these two points.

Fig. 2. A necessary case of horizontal rotation based on two unbalanced tops of nostrils

Let *ep* be a threshold defined as a hard constraint to normalize the face. If the deviation of two tops of nostrils is larger than *ep*, rotation of the face image will be done. Depending on the direction of face deviation (i.e. left or right), the image may be rotated $1°$ or $-1°$ horizontally according to the following Algorithm 1.

Algorithm 1. Horizontal rotation

1: define *ep* threshold
2: **while** (| *leftNostrils.y* – *righNostrils.y* | > ep)
3:    **if** (*leftNostrils.y* > *righNostrils.y*) rotate *image* $1°$ horizontally
4:    **else** rotate *image* $-1°$ horizontally
5:    get *leftNostrils*; get *righNostrils*;
6: **end while**

Vertical Rotation. To conclude vertical deviation of a face, we compare the input data with its quantization form. First a straight line is drawn paralleling to the horizontal coordinate axis in the quantization map (see Fig. 3). Our target is then to rotate the face so that the vertical axis of the face is perpendicular to this line. In this case, top of the nose is chosen as the center to separate its left and right side.

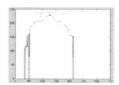

Fig. 3. A quantization map of a face deviated vertically in its right side

Let l be a length of the straight line drawn from the top of the nose to its left and right side. *SoL* and *SoR* are defined as the left and right sides of the top in the quantization map, respectively. Algorithm 2 is used to rotate the face vertically.

Algorithm 2. Vertical rotation

1: define *ep* threshold
2: **while** (| *SoL − SoR* | > *ep*)
3:     **if** (*SoL* > *SoR*) rotate *image* 1° vertically
4:     **else** rotate *image* -1° vertically
5:     *Line* ← *image* [*Nose.x − l* : *Nose.x + l, Nose.y*]
6:     *SoL* ← sum(*Line*[1 : *l*]); *SoR* ← sum(*Line*[*l*+1 : *l*+100]);
7: **end while**

Since the order of vertical and horizontal rotations can affect the result of face normalization, rotating first horizontally and then vertically gains better results than in the reverse order. If we first rotate in vertical direction, the straight line cannot be perpendicular to the vertical axis of the face any more after its horizontal rotation. As a result, its rotated result becomes incorrect. Hence, we decide to perform the horizontal rotation before doing the vertical one.

2.3   ROI Extraction

After normalizing, the face image is very close to a front face. Next we will find biometric features of the face and then apply a local extreme method to determine ROIs in the image. There are four ROIs including two eyes, a nose, and a mouth that we concern. Finally rectangles will enclose each ROI.

As [11], human face can be divided on a golden ratio (see Fig. 4(a)). This ratio is considered as special harmony of the face. Moreover, in 1994 Farkas introduced facial anthropometric measurement points [12], that will transform simultaneously and whose positions are rarely changed during the variation of a face. In this paper, we only choose some of these points to extract ROIs. As shown in Fig. 4(b), these selected points are on a line along the nose.

(a)   (b)

Fig. 4. (a) Golden ratio and (b) some invariant points on a face

By applying a local extreme method into a quantization chart of the line along the nose, we can obtain local extreme points as anthropometric measurement points (see Fig. 5).

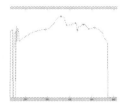

Fig. 5. A quantization chart of the line along the nose

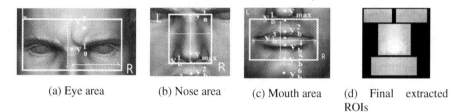

(a) Eye area   (b) Nose area   (c) Mouth area   (d)   Final   extracted
                                                                     ROIs

Fig. 6. Facial ROI extraction

First, an eye area is illustrated in Fig. 6(a) based on three points y_u^3, y_u^1, and y_{max} and described as

$$\left\{ (x, y) \mid x_{(L_3)} \leq x \leq x_{(R_3)}, \left(y_u^1 + y_{max} \right) / 2 \leq y \leq y_u^2 \right\} \tag{1}$$

Next, nose area is illustrated in Fig. 6(b) based on two points y_u^1 and y_b^1 and described as

$$\left\{ (x, y) \mid x_{(L_1)} \leq x \leq x_{(R_1)}, y_b^1 \leq y \leq y_u^1 \right\} \tag{2}$$

Finally, mouth area is illustrated in Fig. 6(c) based on two points y_b^5 and y_b^1 and described as

$$\left\{ (x, y) \mid x_{(L_2)} \leq x \leq x_{(R_2)}, y_b^5 \leq y \leq y_b^1 \right\} \tag{3}$$

Based on the facial golden ratio, the width of rectangles enclosing these eyes, nose, and mouth ROIs are estimated respectively (see Fig. 6(d)) as below

$$\mid x_{(L_3)} - x_{(R_3)} \mid = \frac{\mid y_u^1 - y_b^3 \mid}{\varphi} \tag{4}$$

$$\mid x_{(L_1)} - x_{(R_1)} \mid = \frac{\mid y_{max} - y_b^4 \mid}{\varphi} \tag{5}$$

$$\mid x_{(L_2)} - x_{(R_2)} \mid = \mid y_{max} - y_b^4 \mid \tag{6}$$

2.4   Feature Extraction Using SIFT Descriptor

SIFT descriptors was proposed by Lowe [13] in 2004 and there have been many its improvements [15], [16]. Compared to others, this method has its own advantages of detecting image features at different scales relied on scale space theory [17] as well as being invariant to Affine transformations and illuminative changes. SIFT method processes in three main stages: locating features named as key-points on an image, generating detectors, and generating descriptors. These key-points usually contain distinct characteristics that help improving the efficiency of the matching stage. In this paper, we use an improvement of SIFT descriptors done by Andrea Vedaldi (University of California) [16].

According to Lowe, SIFT descriptor depends on two parameters. They are the number of octave and the number of picture in each octave (numlevels) [13]. If we can choose a good number of octave and numlevels, good features will be found and the identification will be performed better (see Fig. 7).

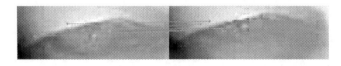

Fig. 7. An illustration of SIFT descriptors at a good level of numlevels for a left eye of the same person captured from two different pose directions

2.5   Matching

In this stage, key-points of two unknown face images are matched to find a set of similar points in these faces. Basically, pairs of similar key points have to locate in similar ROIs in the face (eye to eye, nose to nose, and mouth to mouth). Thus we evaluate the similar location of each pair of two similar key points. Each key point of the pair is calculated its relative distance to the top of the noise. Then, if the difference between these two relative distances is less than a threshold λ, this pair is considered to have similarities in both feature and position. Meanwhile, two similar key points located in the different positions will be removed. By counting the number of pairs of similar key points, we can evaluate the similarity in two faces.

In our proposed method, the given result is only the closest face to the face we want to identify. It cannot answer whether the result is correct or not. As we know, different persons usually have at least one eye that is absolutely dissimilar from the corresponding eye of others (explained later). In few cases, if both eyes have similar key points, the subtraction of the number of similarities key points between two eyes of the same person is lower than of two different persons. Next, we perform two statistics in our database corresponding to two above comments respectively. Following Table 1 shows results of the first statistics. We took arbitrary 10 persons out of 80-person dataset in succession to build more than 800 observations.

For the second statistics, we first call *SL* as the number of similar key points of the left eye and *SR* as the similar points of the right eye between the input face and the current matching face. In the case that a person have similar key points in both his/her eyes, an experiment to measure |*SL* − *SR*| is represented in Fig. 8.

Table 1. A statistics of the first comment about the similarity of key points in human eyes

Numlevels	At least one eye has no similar key point
2	606(75.8%)
3	599(74.9%)
4	594(74.3%)
5	593(74.1%)
6	542(67.8%)
7	511(63.9%)

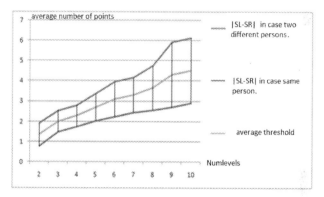

Fig. 8. A statistics of the second comment about the |*SL* − *SR*| value between the same person and different persons

Based on the above statistics, we suggest an adaptive way to identify whether the input face belongs to the given dataset or not. The matching result is not correct (that means it is not posed from the same person) if it satisfies one of the following conditions:

$$SL{=}0 \text{ or } SR{=}0 \qquad (7)$$

$$SL > 0 \text{ and } SR > 0 \text{ and } |SL - SR| \geq \alpha \qquad (8)$$

where α is an average value corresponding to each Numlevel chosen in the green line shown in Fig. 8. Algorithm 3 is used to describe how to match the given input with the existing system database.

3   Results

To demonstrate performance of our proposed 3D human face recognition system, we have carried out different evaluations on the Notre Dame biometric 3D-TEC dataset (including 440 poses of over 90 persons). Our program runs on a PC equipped with Intel Core 2 Duo, CPU 2x2.0GHz, 3GB RAM, Windows 7 Professional in the environment of Matlab 2011.

To evaluate the performance of face recognition, we perform our model in two data collections. They are selected randomly by choosing a quarter and a third of the dataset. Due to the ROI extraction and our remarks on the matching stage, we can both save a lot of cost for the identification and achieve high accuracy of recognition. As shown in Fig. 9, the minimum average cost is approximate 13.2 seconds in Matlab. Furthermore, as described in Table 2, the maximum precision is up to 84.6%.

Algorithm 3. Matching

Input: *inputImage*

1: define α, λ

2: max $\leftarrow$ 0

3: **for all** *imageSet*

4:      $\theta \leftarrow getSIFT(inputImage, imageSet[i]$);

5:      $SL \leftarrow |\theta_{leftEye}|$; $SR \leftarrow |\theta_{rightEye}|$;

6:      $S \leftarrow SL + SR + |\theta_{Nose}| + |\theta_{Mouth}|$;

7:      **if** $(SL=0 \mid SR=0 \mid (SL>0 \& SR>0 \& |SL - SR| \geq \alpha))$

8:              **continue**

9:      **if** $(max < S)$ $max \leftarrow S$; $\omega = i$; **end if**

10: **end for**

11: **if** $(max = 0)$ we cannot find any similar face.

Output: *imageSet*[ω]

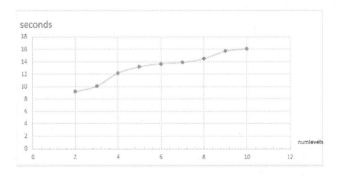

Fig. 9. Average implementation cost for our face recognition model

Table 2. The precision though the numlevel with a quarter of database and one-third of database

Numlevels	Precision with a quarter of dataset (%)	Precision with a third of dataset (%)
2	70.5	63
3	72	65.3
4	75.4	67.2
5	81	68.3
6	84.6	71.3
7	84.3	70.2

In the next experiment, we first took 10 persons out of the database. Face images of these persons are not used to train the system. They are considered as "strangers" in the database. We later use them to test our proposed system. As shown in Fig. 10, we finally get 20% of the minimum False Acceptance Rate (FAR) in the case of choosing good SIFT parameters.

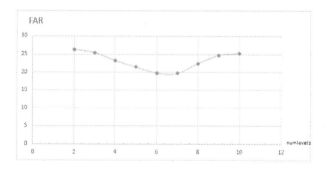

Fig. 10. FAR when testing 10 "strangers" in the training dataset

4   Conclusions

We represented a novel 3D face recognition algorithm to model good human identification in a simple and fast way. Since SIFT descriptors is a suitable method in 2D database, we tried to reduce its high cost of many processing steps as well as limit the dependence on its own parameters for applying into 3D data. In the experiments, our proposed model showed promising results of efficiency and effectiveness. However, this algorithm still needs further improvements. Although its precision and average cost is acceptable, they have not reached the required criteria of a security system yet. And appropriate parameters for SIFT descriptors should be considered more adaptively to restrict coincidental positions of key points.

Acknowledgement. Thanks to UND Principal Investigator for permission to use Notre Dame biometric 3D-TEC.

References

1. Bowyer, K.W., Chang, K., Flynn, P.: A survey of approaches and challenges in 3D and multi-modal 3D + 2D face recognition. Journal of Computer Vision and Image Understanding 101(1), 1–15 (2006)
2. Wang, Y., Chua, C., Ho, Y.: Facial feature detection and face recognition from 2D and 3D images. Pattern Recognition Letters 23, 1191–1202 (2002)
3. Pan, G., Han, S., Wu, Z., Wang, Y.: 3D Face Recognition using Mapped Depth Images. In: IEEE Computer Society Conference on CVPR (CVPR 2005) Workshops, vol. 03, p. 175 (2005)
4. Yuan, X., Lu, J., Yahagi, T.: A Method of 3D Face Based on Principal Component Analysis Algorithm. IEEE International Recognition Symposium on Circuits and Systems 4, 3211–3214 (2005)
5. Russ, T., Boehnen, C., Peters, T.: 3D Face Recognition Using 3D Alignment for PCA. In: IEEE Computer Society Conference on CVPR (CVPR 2006), vol. 2, pp. 1391–1398 (2006)
6. Taghizadegan, Y., Ghassemian, H., Naser-Moghaddasi, M.: 3D Face Recognition Method Using 2DPCA-Euclidean Distance Classification. ACEEE International Journal on Control System and Instrumentation 3 (2012)
7. Gervei, O., Ayatollahi, A., Gervei, N.: 3D Face Recognition Using Modified PCA Methods. World Academy of Science, Engineering & Technology 4, 254–257 (2010)
8. Cook, J., McCool, C., Chandran, V., Sridharan, S.: Combined 2D / 3D Face Recognition using Log-Gabor Templates. In: IEEE International Conference on Video and Signal Based Surveillance (AVSS 2006), p. 83 (2006)
9. Pamplona Segundo, M., Silva, L., Bellon, O.R.P.: Real-time scale-invariant face detection on range images. In: IEEE International Conference on Systems, Man and Cybernetics (SMC), pp. 914–919 (2011)
10. Jahanbin, S., Cho, H., Bovik, A.C.: Passive Multimodal 2-D+3-D Face Recognition Using Gabor Features and Landmark Distances. IEEE Transactions on Information Forensics and Security 6, 1287–1304 (2011)
11. Rossetti, A., De Menezes, M., Rosati, R., Ferrario, V.F., Sforza, C.: The role of the golden proportion in the evaluation of facial esthetics. Angle Orthodontist 83 (2013)
12. Farkas, L.G.: Anthropometry of the head and face. Raven Press, New York (1994)
13. Lowe, D.G.: Object recognition from local scale-invariant features. In: IEEE International Conference on Computer Vision, vol. 2, pp. 1150–1157 (1999)
14. Lindeberg, T.: Scale-space theory: A basic tool for analysing structures at different scales. Journal of Applied Statistics 21(2), 224–270 (1994)
15. Lowe, D.G.: Distinctive image features from scale-invariant keypoints. International Journal of Computer Vision 60(2), 91–110 (2004)
16. http://www.vlfeat.org/~vedaldi/code/sift.html
17. Lindeberg, T.: Scale-space theory: A basic tool for analysing structures at different scales. Journal of Applied Statistics 21, 224–270 (1994)

The Bootstrap Approach to the Comparison of Two Methods Applied to the Evaluation of the Growth Index in the Analysis of the Digital X-ray Image of a Bone Regenerate

Aneta Gądek-Moszczak[1], Jacek Pietraszek[1,*], Barbara Jasiewicz[2],
Sylwia Sikorska[1], and Leszek Wojnar[1]

[1] Institute of Applied Informatics, Cracow University of Technology,
Al. Jana Pawla II 37, 31-864 Kraków, Poland
{aneta.moszczak,pmpietra,leszek.wojnar,ssikorska}@gmail.pl
[2] Jagiellonian University, University Orthopedics-Rehabilitation Hospital,
ul. Balzera 15, 43-501 Zakopane, Poland
basiaj@klinika.net.pl

Abstract. The bone elongation by Ilizarov method is the important way of treatment of physical disability. During the elongation is necessary to control the accumulation of a regenerate and indicate the moment when fixator should be removed. The control is performed by observing the X-ray images and evaluation of the Regenerate Development coefficient. The automated image analysis is used to ensure the objectivity and reproducibility of the measurement. The authors have developed a new method for evaluating coefficient reflecting the specific treatment regenerate parts of an image associated with intramedullary nail. To assess the equivalence of the method with respect to the current, the bootstrap method was applied to sample of 23 X-ray images. The bootstrap method provided the simulated distribution of the difference between the index obtained from the new method and the index obtained from the previous one.

Keywords: quantitative regenerate assessment, image processing, bone maturity, Ilizarov method, bootstrap, statistical analysis.

1 Introduction

The process of limbs lengthening is a long and painful procedure which results in correcting the limbs disproportion. The limbs length disproportion can be an inborn defect, the result of an earlier disease, a defective bone union or a bone defect. An uneven length of limbs can cause serious disorders in the whole loco-motor systems. Surgical intervention, especially in the case of large disproportion, is the only one possibility to decrease the limbs length disproportion. The history of limbs lengthening was started by Codivilli, who performed the elongation of a femoral bone in 1905

---

* Corresponding author.

© Springer International Publishing Switzerland 2015
D. Camacho et al. (eds.), *New Trends in Computational Collective Intelligence*,
Studies in Computational Intelligence 572, DOI: 10.1007/978-3-319-10774-5_12

[1]. Numerous methods of limb lengthening have been developed and are applied in clinical practice [2]. Depending on the method, various types of a stabilizer are used to stabilize the elongated limb. Stabilizers can by divided into two types: external and internal. Ilizarov frame fixator is an example of external fixators, whereas the intramedullary nail is used as an internal fixator [3].

One of the most popular techniques of limb lengthening is Ilizarov method [1, 4]. In this method, an external fixator invented by Ilizarov is used. This fixator is made up of circular rims, threaded rigid or telescopic spacer rods and Kirscher arrow-heads (Fig.1). The construction of the fixator was designed to ensure mutually coaxial positioning of the bone fragments, which ensures the correct union of the bone. In the process of elongation, the periscopic rods allow for the precise adjustment of the distraction space. The fixator also takes the load off the area of the bone distraction, ensuring the necessary rigidity of the limb.

 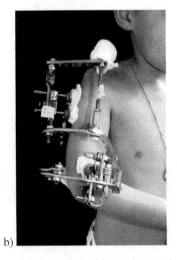

a)   b)

Fig. 1. Patients with Ilizarov frame fixators in the course of elongation of femoral (a) and forearm bone (b)

The Ilizarov frame fixator is well-proven and widely applied. But currently, to improve the patient's comfort and reduce complication during the process of elongation, internal lengthening fixators are being developed. The Intramedullary Skeletal Kinect Distractor (ISKD) is an example of an internal fixator that is inserted inside the intramedullary canal after osteotomy [5]. An internal fixator makes the distraction process more comfortable for patients, because it doesn't diminish the comfort of living. The ISKD consists of two telescope cylinders connected by a threaded rod (Fig. 2). The junction between these two cylinders is surrounded by a keyring collar which secures the connection and prevents the nail from turning more than it is required for lengthening. The lengthening is performed by two one-way clutches that create an unidirectional drive mechanism through oscillating rotation between 3 and 9°. To achieve 1 mm of elongation, at least 60 rotations are needed, which are obtained by everyday movements. The lengthening is achieved with no additional surgeon's or patient's intervention, but just by patient's everyday regular activity. Some details describing a method as well as risk factors may be found in papers [6-9]. The limb elongation process can be divided into several stages. The first operative stage consists in installing the frame fixator and performing coricotomy. The second stage is elongation. The elongation rate is 1 mm per day in 4 steps of 25 mm each. The final stage of the treatment is the stabilization of the new bone tissue, called regenerate [10]. At the stage of stabilization of the regenerate, gradual mineralization takes place [1]. When the level of the regenerate mineralization reaches the level of that of a healthy bone, it is assumed that it is capable of being loaded, and the patient with no risk can resume completely normal life. To control the correctness of the elongation, and after that of the stabilization process, the X-ray images are taken within the period of 3 or 4 weeks. Having visually assessed the progress of the treatment, the surgeons can make a diagnosis and adjust the treatment to prevent the possible complications.

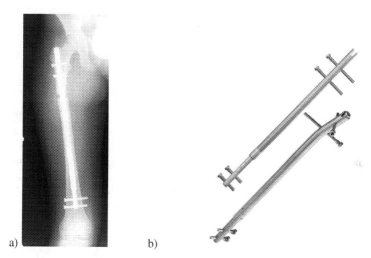

a)   b)

Fig. 2. X ray of the limb elongated by ISKD (a), Intramedullary Skeletal Kinect Distractor (b) (http://intl.orthofix.com)

The radiological examination is a basic test used in everyday clinical practice. The X-ray images are taken in two projections: anterior-posterior (AP) and lateral (L). In recent years, densitometric, ultrasonographic and tomographic techniques are also used to assess the changes in the new bone structure, but these are only complementary methods in relation to the X-ray examination.

A crucial moment in the whole stabilization process is the moment of removing the fixator. If the fixator is removed too early, when the new bone tissue has not reached the optimal mineralization level to be able to load the body weight, it can cause the regenerate bending and its deformation, which is painful and forces the patient to undergo additional treatment. Removing the fixator too late can cause the bone break because of decalcification of the host bone segment, and what is more, it causes long-lasting discomfort for the patient. As it was already mentioned, all diagnosis decisions are taken by surgeons, mainly on the base of a visual analysis of the X-ray images and standard patient examination.

That method is highly subjective and strongly depends on the experience of the surgeon and the quality of the radiographs. In the face of the risk of further complication, there is a need to develop an objective method that could give more accurate information about the process of calcification of the new bone tissue.

There is no standard method for the quantitative assessment of the regenerate quality. Cope and Samchukov [11] proposed a method of evaluation of changes in the regenerated bone tissue in the course of mandibular bone elongation. On the X-ray images they tried to determine the geometric parameters of the regenerate, its density, and the presence or absence of the bone cortex. As a result, they obtained the so-called aluminum volume equivalent which provides indirect information on the degree of calcification of the analyzed tissue. However, the method never progressed beyond the experimental stage. Dinah [12] tried to estimate the time required by the bone regenerate tissue to form the cortex. He based his research on the statistical analysis and on the relation between the index of regenerate growth and the value of bone lengthening. That method is burdened with a large margin of error, and in case of complications in the regeneration process, completely useless. Another attempt of the quantitative assessment of the bone regenerate was presented by Hazra et al. [13]. They proposed to calculate the pixel value ratio of the regenerate region and the proximal area, and for the same region, to indicate the ratio of BMD vaules (Bone Mass density), measured by the DEXA technique. Next, the pixel value ratio and the BMD ratio were compared, using Pearson correlation coefficient, and obtained quite a good correlation (from 0.85 to 0.91 for all tested pairs). Another method, which was devised by one of the authors [14, 15], proposed the so called Coefficient of Regenerate Development (CRD), calculated by the quantitative comparison of the real bone structure and the model structure. The method of the CRD assessment was limited only to Ilizarov method. The possibilities of extending the method's application to the elongation process with the ISKD fixator have been verified. The method is presented in detail in the Materials and Method subsection of this paper.

2   Materials and Method

The conventional X-ray images of limbs elongated according to Ilizarov method at the Clinic of Orthopedic and Rehabilitation, Collegium Medicum, Jagiellonian University were used as the research material for the study. All analyzed images are part of the standard documentation of the patient's treatment. All images were digitalized with the resolution of 254 dpi, which corresponds to 0.1 mm of the spatial resolution. The parameters of digitalization correspond to the resolution of photosensitive films on which the X-ray images were formed. A digital representation of the X-ray images was saved as JPEG graphics files with a low compression ratio. It was experimentally established that saving an image in a JPEG format does not entail any significant loss of information and allows for approximately 10-fold reduction of the file size. Image processing and the analysis were performed using Aphelion software and the algorithm worked out by the authors [15].

The images selected for the analysis present the history of the elongation process of 5 patients, which provides 50 images for the analysis (Fig.3).

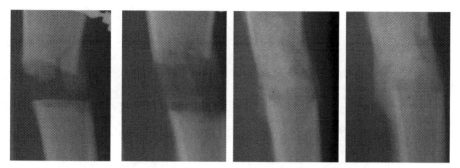

Fig. 3. An example of X-ray images illustrating bone tissue regeneration process

The aim of the experiment was to verify whether the method proposed by Gądek [14, 15] may be used to assess the bone regenerate maturity when the elongation is performed with the ISKD. The method of the Coefficient of Regenerate Development indication is based on the ratio calculated between the ordinary image, presenting the regenerate, and the model image, generated by the algorithm (Fig.4). The model image simulates the healthy bone structure. The image of the regenerate is compared to the model image, simulating healthy bone structure. The model image is interpolated by the linear function on the base of the value of the bone under and below the distraction space.

The next step was to divide the two images of the bone regenerate – the real image and the model image. The arithmetic mean of the pixel value on the final image gives the value of the CRD. One of the most important assumptions of the method was that the central part of the regenerate is analyzed, which is indicated interactively by the operator who draws the rectangular ROI (Region of Interest) (Fig.4).

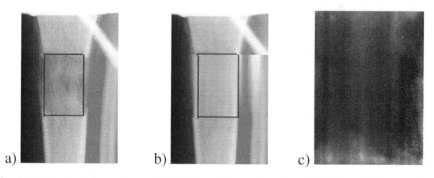

a)   b)   c)

Fig. 4. CRM calculations stages a) indication of the region of analysis; b) model image calculation c) the result of division of the images (a) and (b)

The CRD value close to 1 informs surgeons that the level of calcification of the regenerate is equal to the level of calcification of the host bone segments. As it was mentioned above, this approach concentrates on the central part of the regenerate, and in fact, on the whole regenerate, visible on the image. In the ISKD method, the central part of the regenerate is hidden by the fixator, and part of the information will be covered by the fixator. The authors decided to compare the value of the CRD of images where the regenerate is visible, and on the same images simulate the presence of the ISKD and calculate the SRD value for two regions on both sides of the ISKD (Fig.5).

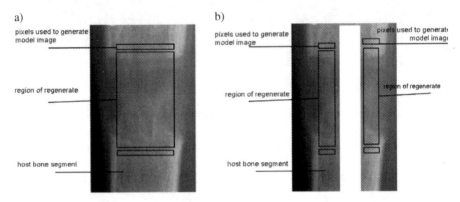

Fig. 5. The regions of the analysis on the X-ray image in Ilizarov method (a) and adaptation to the ISKD method (b)

The analysis of the original images from limb elongation by Ilizarov method was conducted, and next, all analyzed images were changed (Fig. 5b) by adding the white rectangle in the center of the regenerate image, along the bone axes, simulating the ISKD. The analysis process had to be slightly modified because for each image the two regions of the regenerate were analyzed, and for each image 2 values of CRD were provided. The final value of CRD for these images was the arithmetic mean of the value for the right and left regions of the regenerate.

3   Image Analysis Results

All images selected for the analysis were analyzed and the results were compared. The analysis of the correlation between these two groups of the results have shown an acceptable value of the correlation coefficients, which depends on the quality of the analyzed images and the size of the analyzed area. The smaller the area of analysis, the lower the correlation between the calculated values of the regeneration factor for the entire area of the regenerate and simulations of elongation by ISKD method.

Table 1. Data obtained from Ilizarov and ISKD methods

Patient	Projection	Data ID	Ilizarov method (M1)	ISKD method (M2)	Difference (D)
1	AP	4.01. 02	0.75	0.74	0.015
		4.02.02	0.70	0.63	0.075
		20.05.02	0.92	0.84	0.08
		3.07 02	0.97	1.04	-0.07
	L	20.05.02	0.88	0.85	0.03
		3.07.02	0.91	1.10	-0.19
2	AP	6.11.01	1.28	0.95	0.33
		15.01.02	0.75	0.55	0.2
		19.02.02	0.63	0.55	0.08
		7.05.02	0.98	0.67	0.31
		1.07.02	0.93	0.73	0.205
	L	7.05.02	0.93	0.99	-0.06
		24.06.02	0.93	0.93	0.005
		1.07.02	0.96	1	-0.04
		13.08.02	0.97	1.06	-0.09
3	AP	27.09.01	0.9	0.87	0.03
		6.11.01	0.92	0.87	0.05
		4.12.01	0.93	0.89	0.04
		19.02.02	0.97	0.925	0.045
		9.04.02	1.02	0.94	0.08
	L	27.09.01	0.66	0.52	0.14
		9.10.01	0.69	0.65	0.045
		6.11.01	0.85	0.84	0.015
		4 12.01	0.91	0.88	0.035
		18.01.02	0.93	0.87	0.065
		19.02.02	0.96	0.99	-0.035
4	AP	20.11.01	0.82	0.81	0.01
		7.01.02	0.94	0.91	0.03
		5.03.02	0.96	0.93	0.03
		14.05.02	1.01	0.97	0.04
		25.06.02	1.02	0.95	0.07
	L	4.09.01	0.32	0.36	-0.035
		26.09.01	0.43	0.49	-0.055
		20.11.01	0.57	0.72	-0.15
		7.01.02	0.85	0.74	0.115
		5.03.02	0.97	0.93	0.04

Table 1. (*continued*)

Patient	Projection	Data ID	Ilizarov method (M1)	ISKD method (M2)	Difference (D)
5	AP	20.11.01	0.87	0.87	0
		26.02.02	0.91	0.98	-0.065
	L	5.06.01	0.95	0.46	0.49
		6.03.07	0.57	0.605	-0.035
		4.01.02	0.74	0.82	-0.08
		26.03.14	0.9	0.845	0.055
		19.05.02	0.82	0.965	-0.145
		7.05.03	0.94	1	-0.06
		8.03.08	0.92	1	-0.08

4   Statistical Analysis

The paired results obtained from the Ilizarov (M1) and ISKD (M2) methods are arithmetically different, which is obviously due to the subtle details of images selection. The question which should be posed is whether the differences between the methods are significantly different from zero or not. The results (Tab.1.) may be treated as the realization of the random variables M1 and M2 over the experimental object (combination of the patient and the projection). Such an approach leads to the formal definition of the methods' difference, stated as:

$$D = M1 - M2 \tag{1}$$

where D is the random variable with unknown distribution. The question about the statistical significance of the non-zero difference between the considered methods may be replaced by a typical statistical hypothesis [16]:

$$H_0: \text{mean}(D) = 0 \tag{2}$$

with an alternative hypothesis:

$$H_1: \text{mean}(D) \neq 0. \tag{3}$$

The analytical testing of the hypothesis H0 at the significance level α may be changed into the equivalent, but a more practical, approach [17]: testing whether the confidence interval of the mean of the random variable D at $(1-\alpha)$ confidence level contains the zero value or not. Due to the fact that the data set is rather small (only 45 values), the bootstrap method [18] appears to be the proper approach to identify numerically the distribution of the random variable D and to evaluate the bounds of the confidence interval of the mean.

The bootstrap draw was set to 100,000 times with the population size of 45 values to facilitate the identification of the cumulative probability of the quantile with the zero value. The obtained bootstrapped histogram of the mean (D) is shown in Fig.6. with the position of the zero value specially displayed by the arrow sign.

The mean value of the random variable D is 0.03477 with 95% confidence interval between 0.00056 and 0.07256. It means that the zero value is outside the confidence interval and the alternative hypothesis H_1 should be selected instead of the rejected null hypothesis H_0. Such a decision is also confirmed by the fact that the zero value has assigned the cumulative probability at 0.0226, which is below the critical 0.025. More practically, it means that the results obtained from Ilizarov and ISKD methods differ significantly and cannot be treated equivalently.

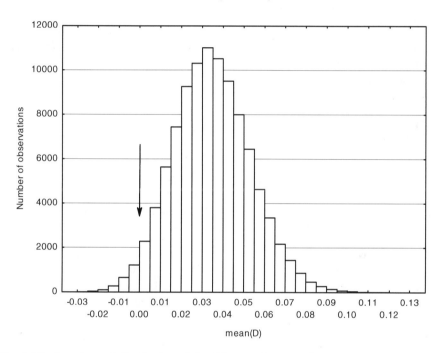

Fig. 6. The bootstrapped distribution of the difference between the methods of Ilizarow and ISKD. The bootstrap consists of 100,000 draws from the population of 45 values. The arrow shows the position of the zero value.

5   Conclusion

Developing methods for the quantitative assessment of the maturity or quality of bone regenerate tissue is important and worth further study. Objectivity and sensitiveness to even slight changes on X-ray images of the bone regenerate may help surgeons to make appropriate treatment decisions, which in consequence, may cause less complication and less pain for patients. In the presented case, the considered methods are not simply replaceable but it still is unknown if the observed bias may be known a priori. In further studies, the authors will concentrate on working out the correction coefficient and on analyzing the X-ray images of the limbs elongated with ISKD.

References

1. Paley, D.: Problems, Obstacles, and Complications of Limb Lengthening by the Ilizarov Technique. Clin. Orthop. Relat. R 81–104 (1990)
2. Paley, D.: History and Science Behind the Six-Axis Correction External Fixation Devices in Orthopaedic Surgery. Operative Techniques in Orthopaedics 21, 125–128 (2011)
3. Dobyns, J.H., Wood, V.E., Bayne, L.G.: Congenital hand deformities. In: Green, D.P. (ed.) Operative Hand Surgery, Churchill Livingstone, New York (1993)
4. Babatunde, O.M., Fragomen, A.T., Rozbruch, S.R.: Noninvasive quantitative assessment of bone healing after distraction osteogenesis. HSS Journal: the Musculoskeletal Journal of Hospital for Special Surgery 6, 71–78 (2010)
5. Cole, J.D., Justin, D., Kasparis, T., DeVlught, D., Knobloch, C.: The intramedullary skeletal kinetic distractor (ISKD): first clinical results of a new intramedullary nail for lengthening of the femur and tibia. Injury 32, 129–139 (2001)
6. Kenawey, M., Krettek, C., Liodakis, E., Meller, R., Hankemeier, S.: Insufficient bone regenerate after intramedullary femoral lengthening risk factors and classification system. Clin. Orthop. Rel. Res. 469, 264–273 (2011)
7. Mazeau, P., Assi, C., Louahem, D., L'Kaissi, M., Delpont, M., Cottalorda, J.: Complications of Albizzia femoral lengthening nail: an analysis of 36 cases. J. Pediatr. Orthop. – Part B 21, 394–399 (2012)
8. Wang, K., Edwards, E.: Intramedullary skeletal kinetic distractor in the treatment of leg length discrepancy – a review of 16 cases and analysis of complications. J. Orthop. Trauma. 26, E138–E144 (2012)
9. Yamaguchi, K., Yanagimoto, S., Kageyama, T., Fujita, Y., Funayama, A., Kanaji, A.,, S.M., Toyama, Y.: Experimental assessment of a novel intramedullary nail for callus distraction by the segmental bone transport method. J. Orthop. Sci. 19, 323–331 (2014)
10. Eyres, K.S., Bell, M.J., Kanis, J.A.: New bone formation during leg lengthening. Evaluated by dual energy X-ray absorptiometry. J. Bone Joint Surg. Br. 75, 96–106 (1993)
11. Cope, J.B., Samchukov, M.L.: Mineralization dynamics of regenerate bone during mandibular osteodistraction. Int. J. Oral Max. Surg. 30, 234–242 (2001)
12. Dinah, A.F.: Predicting duration of Ilizarov frame treatment for tibial lengthening. Bone 34, 845–848 (2004)
13. Hazra, S., Song, H.R., Biswal, S., Lee, S.H., Lee, S.H., Jang, K.M., Modi, H.N.: Quantitative assessment of mineralization in distraction osteogenesis. Skeletal Radiology 37, 843–847 (2008)
14. Gądek, A., Wojnar, L., Tęsiorowski, M., Jasiewicz, B.: A new method for quantification of regenerated bone tissue on X-ray images of elongated bones. Image Analysis and Stereology 22, 183–191 (2003)
15. Gądek, A.: Computer image analysis of bone regenerate in the Ilizarov's method. Institute of Applied Informatics, Cracow Univ. of Technology, Cracow (2005)
16. Rohatgi, V.K., Saleh, A.K.M.E.: An introduction to probability and statistics. John Wiley & Sons, New York (2001)
17. Chernick, M.R.: Bootstrap Methods: A Guide for Practitioners and Researchers. John Wiley & Sons, Hoboken (2008)
18. Shao, J., Tu, D.: The Jackknife and Bootstrap. Springer Science+Business Media, New York (1995)

Part III
Computer Vision Techniques

Storytelling of Collaborative Learning System on Augmented Reality

Seung-Bo Park[1], Jason J. Jung[2], and EunSoon You[1]

[1] Institute of Media Content, Dankook University, South Korea
[2] Dept. of Computer Engineering, Yeungnam University, South Korea
{Molaal,tesniere}@naver.com, j2jung@gmail.com

Abstract. As augmented reality technologies offer superior senses of presence and immersion to learners through layers of virtual information over the physical world, much attention has been drawn to this learning medium that creates a new learning environment driven by experiences. And, existing AR-based learning content has just focused on interactions among users and on three-dimensional imaging techniques while the collaborative element has not been taken into account fully. Against the backdrop, a rule-based interactive storytelling element has been introduced to the system that offers a variety of contents at a user's choice so as to draw attention and interests. To verify the effectiveness of such learning system, an experiment was conducted with primary school students who went through a performance assessment of 'Growing Peach Tree'. The experiment proved the learning system effective with a high score of 80.4, which is equivalent to that of debate-based learning.

Keywords: Storytelling, collaborative learning, augmented reality, rule-based, smart device.

1 Introduction

The learning environment using information and communication technologies has continued to evolve from e-learning enabled by a computer to u-learning in the ubiquitous environment and to smart learning that utilizes smart devices. Of late, the augmented reality technologies have come into focus as a learning medium that can create a new learning environment driven by experiences. The augmented reality allows virtual objects to blend into the physical world in real time, which has been introduced to various areas of defense, health, construction and game among others since the 1990s.

A significant number of researchers have pointed out that collaborative learning is where the greatest potential of the AR lies [1-4]. Shelton and Hedley developed AR-based content regarding the Earth's revolution and rotation by using ARToolKit, which is an open source library for AR application programs, and used the contents in class [5]. A tool for 3D geometry, Construct3D, was developed based on a mobile collaborative AR system, Studierstube, enabling a teacher and students to collaborate face-to-face [1,6]. MagicBook uses an actual book as an interface for a user to switch

© Springer International Publishing Switzerland 2015
D. Camacho et al. (eds.), *New Trends in Computational Collective Intelligence*,
Studies in Computational Intelligence 572, DOI: 10.1007/978-3-319-10774-5_13

between physical and virtual worlds freely [7,8]. The user controls a 3D virtual object appearing above the open book through a head mounted display or HMD.

Similar to the study is experiential AR Gardening system [9] and Narrative-based Immersive Collaborative Environment (NICE) [10] created for biology education. The AR Gardening takes account of physiological and environmental elements needed to grow a plant, thereby providing an environment where users naturally acquire diverse background knowledge that affects the gardening. Furthermore, a user may water a virtual flower or bask it in the sunshine while experiencing how the flower grows over the time depending on interactions that take place along the way. However, the AR Gardening does not take the collaborative element into consideration and presents a challenge to the AR learning agent required to accurately understand various physical and emotional responses exhibited by users. NICE is an AR system that enables users to have a virtual garden through collaboration, where they can control clouds and the sun to grow flowers while sharing the virtual garden. This system is rooted in the constructivism learning theory, which explains how a learner generates knowledge and constructs understanding on their own, and produces narrative-based learning content that encourages users to create a story inspired by various interactions exchanged between a user and the garden.

There was a case study that suggested storytelling as a means to facilitate user collaboration in the AR environment [11]. Storytelling is a compound word made up of "story" and "telling", which refers to a method of narrating a story. In the field of pedagogy, story-driven contextual enrichment is already being considered as an essential step to generate more learning results by encouraging users to participate in the learning process proactively and to become aware of the given learning environment [12-15]. Braun points out the lack of collaborative AR systems introducing the storytelling element and comments that users have fun and pleasant experiences through storytelling and thus are encouraged to collaborate more with others [11]. Interactive storytelling has received considerable attention of late since it allows users to interact with a story through a narrative structure, within which a user may affect the narrative development or change the story, in contrast to the traditional storytelling narrated just by the writer.

Investigation of existing AR learning contents led to discover a limitation which elements of collaboration and storytelling have not been fully considered. Despite the fact that collaboration and storytelling are crucial in generating more learning results, existing studies have given more focus to interactions occurring between users and contents and to visual elements, such as computer graphics.

With an objective to redress the aforementioned issues, this study devised a storytelling-based mobile AR system for collaborative learning and introduced it to primary school students 'Growing Peach Tree' to assess their performance. To this end, we developed a collaborative system supporting team organization and facilitating communication among team members while establishing a rule-based causal system driven by interactive storytelling. Students who used the 'Growing Peach Tree' application were asked to make evaluations of the user experience and system functionality to assess how effective the system was in terms of learning.

The structure of this paper is as follows: Chapter Two describes theoretical background of collaborative AR learning and interactive storytelling while Chapter Three introduces a storytelling-based collaborative AR learning system. Chapter Four suggests how a collaborative learning system is established while offering the evaluation of the system performance. Then, Chapter Five presents conclusion and future research direction.

2   Collaborative AR Learning and Interactive Storytelling

Storytelling is unique in that it delivers information and knowledge while keeping receivers interested and impressed. This is why it has been drawing much attention as a new teaching strategy that makes teaching more effective. Many researchers have long been placing emphasis on the element of story as an essential learning tool since it can encapsulate information, knowledge and context into one compact package [16]. Storytelling conveys meaning through a series of specific events, and this is exactly how students spontaneously acquire knowledge through specific situation and context if storytelling is applied to learning. A story is driven by a world comprised of objects and characters, such as protagonist and antagonist, and by events that construct a plot. Such basic elements of a story deliver complex knowledge with ease while stimulating curiosity and interests. Such story-driven curiosity and interests facilitate collaboration among students while immersing them more in learning [11,19].

Unlike the traditional storytelling that narrates a story through a linear structure of beginning-middle-end led by a writer, the interactive storytelling that focuses on the interactive aspect promoted by the digital media follows a non-linear structure, through which a story can be narrated in many different directions depending on choices made by users. Such interactive storytelling has been applied to computer games that take a story in many different directions through interactions exchanged among users so as to maximize immersion and interests. This study also employs the non-linear narrative structure, which is one of the components that comprise the interactive storytelling, in order to process actions and choices made by users in diverse environments and conditions.

3   Storytelling Design of Collaborative Learning System in the AR

3.1   System Overview

The system is consisted of Client and Server as illustrated in Figure 1. The Client is installed and activated once an application is downloaded for the smart phone, and the Server is comprised of Collaboration Manger and Content Manager. The Client consists of UI Module, AR Module, and Information Retrieval Module for users to manage content and communicate with other users. When a user detects a marker with a phone-embedded camera, the Client then retrieves target content from the

Server to be augmented onto the marker. In addition, a user navigates an input controller for interactions with content. The Input Controller offers a variety of input buttons for content growth as well as functions, such as Messenger and invitation button, for social interactions among users. The Output Module displays content to be superimposed over a target marker. A function has been added for users to search information related to content over the internet.

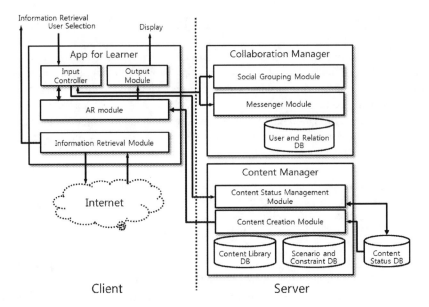

Fig. 1. System Overview

The Server consists of Collaboration manager and Content manager. The collaboration manager handles a group of users participating in learning while supporting interactions among users. The Content Manager manages content status shared by a user group while operating a Content Status Management Module that updates content in line with input provided by users, designed scenarios and constraints and a Content Creation Module that generates content assisted by a Content Library DB and that sets conditions for content growth. In addition, there are Content Status DB that stores content over the course of content growth and Scenario and Constraint DB that manages content growth in line with established stories. Moreover, there is a Content Library DB made up of exterior factors involved with content.

3.2   System Architecture for Storytelling

For collaborative learning content to embrace the element of storytelling, content shall respond to and be altered by user activity. The system installed for content shall be able to process interactive storytelling that is structured to have an open-ended scenario. Therefore, user activity just alters immediate status of content while the final outcome of content is determined only by cumulative user activity. Moreover, results driven by user activity shall be different depending on respective environment of content. For instance, positive results are generated if a user waters the tree content when the soil is dry. But, the results would be negative when the soil is moist. In short, diverse activities performed by a user to grow a tree, such as watering, fertilizing, debugging and pruning, lead to the tree's fruition while unsuitable user activities might rather result in decreased fruition or even death of the tree. Furthermore, a crucial storytelling element of character theory is introduced to engage students more while keeping them interested.

One of the essential elements that make up a story is character, which is mainly divided into protagonist, antagonist and tritagonist [17,18]. For example, the system for tree growing presents characters, such as tree, bug, branch, water and fertilizer. Of the characters, the tree is the target for growth and thus assumes a protagonist role. Bugs and branches are considered to be negative factors that disturb the tree's growth. They are reckoned to be antagonists since they affect the tree's growth depending on their timely elimination. On the other hand, water and fertilizer are main drivers of the tree's growth and mostly generate positive impact. Still, they are deemed to be tritagonist since they can also present negative impact on the tree once used in large quantity.

Table 1. Classification and properties of characters

Class	Object	Property	Type	Description
Protago nist	Peach tree	kindOfBugToGet	Integer set	IDs of bug to appear at tree
		quantityOfWaterIntake	Integer	The ration of water intake per an hour
		numOfProperBranch	Integer	The number of branches for best fruition ratio
Antago nist	Bug	killingMaterial	Integer	ID of material or pesticides to be able to kill bug
		timeIntervalForAppearing	Integer	Time interval to appearing
		timeIntervalToActivate	Integer	Time required till a bug is active
	Branch	timeIntervalForAppearing	Integer	Time interval to appearing
Tritago nist	Water	volume	Float	Amount of water
	Fertilizer	amountOfNourishment	Float	Amount of nourishment

In this respect, the storytelling system intended for collaborative learning content shall be designed to first apply the character concept to objects or inputs controlled by a user then to make categorization thereof and shall be developed as a rule-based causal system that can process user activity in diverse settings. To this end, we introduced the character concept for inputs and objects processed by the system. For instance, characters are given as explained in Table 1 when the system is applied to the task of growing the peach tree. The peach tree represents a main character for the storytelling and assumes a protagonist role as the growth target. Antagonists are bugs and branches with properties motioned in Table 1. At a certain time (timeIntervalToActivate) after a bug is discovered, the peach tree is negatively affected with decreased fruition. To kill the bug, killingMaterial can be applied at pesticide or substances that can be used. A branch is added to the tree at a certain time (timeIntervalForAppearing). Water and fertilizer are tritagonists that assist the protagonist. Their properties are limited to just the amount since they generate positive impact of driving the tree's growth in proportion to their used quantity.

There are four kinds of jobs that can be performed by a user, which are watering, fertilizing, pruning and debugging. Any job-related factor that affects the results depending on respective situation is defined as rules. Such rules are described in Table 2 in line with the sample of growing the peach tree. Table 2 is limited to rules applied to bugs when growing the tree. The bug is a factor that brings down the tree's fruition and thus either increases or decreases the fruition to some extent depending on respective situation.

Table 2. Rule Set for Bug of 'Growing Peach Tree'

Rules	Description
IF (A Bug is kept for timeIntervalToActivate at Tree) THEN Fruition of Tree is decreased 3%	This rule is activated when a bug arrives to first attack time. It decreases the fruition of tree for 3%
IF (A Bug is kept for timeIntervalToActivate after First Attack) THEN Fruition of Tree is decreased 1%	This rule is activated when a bug arrives to next attack time after first attack. It decreases the fruition of tree for 1%
IF (A Bug is deleted in timeIntervalToActivate from Tree) THEN Fruition of Tree is increased 7%	This rule is activated when a bug is deleted in time. It increases the fruition of tree for 7%
IF (Used Debugging Material has killingMaterial of Bug) THEN A Bug is killed	This rule is activated when a button of debugging is selected.

4   Implementation and Evaluation

4.1   System Implementation

For system implementation, a smart phone application and a server system were developed. The development of the system was aimed at collaborative learning for building up contents of 'Growing Peach Tree'. The client system has been designed to be implemented and to operate in the Android environment while it is the Linux environment as for the server. For the client to operate properly, 3D acceleration performance shall be superior [20]. Therefore, there is a need to use a smart phone installed with Android 4.0.4 (ICS) version and above. As for development and testing, Pantech Vega Racer was used. And, Java-based Eclipse was used to develop an application for the client. Eclipse's STS version was used to develop the server through Spring framework in the Linux environment. For testing, an application named Post Man that runs in the Chrome browser was used so as to test a RESTFull-based server.

Screens that are displayed and activated in this manner were captured to be shown in Figure 2. Once a marker is recognized, a display screen of the smart phone's application projects a peach tree retrieved for the marker in question.

As shown in Figure 2, a task of 'Growing Peach Tree' was selected to be developed as learning content. A group leader invites potential team members, and a messenger is provided for team members to be able to communicate with others. Users go through five phases of growing a peach tree from a seed phase to a fruition phase(seed, seedling, branch growth, flower, fruit), throughout which four jobs(watering, fertilizing, pruning, debugging) are allocated to and shared by learners who assume such tasks through collaboration so as to satisfy the given conditions.

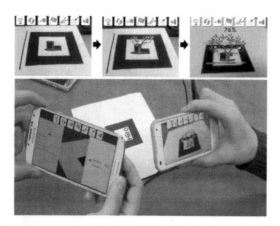

Fig. 2. A Screen Shot for System of 'Growing Peach Tree'

Students share responsibilities for growing the tree and are encouraged to conduct the tasks through collaboration. If a certain input that violates any given condition is entered, a tree is designed to die or to enjoy less fruition. The system is designed to encourage students to seek conditions that are conducive to growing the tree once such an issue is presented.

4.2   Evaluation

User evaluation was conducted on the developed application of 'Growing Peach Tree'. Thirty eight users were divided into ten teams comprised of three to four. Each team was assigned with a goal of growing a tree and asked to use the application to produce fruits by freely exchanging opinions among members. A questionnaire was used to evaluate learning results enjoyed by teams that succeeded in fruition and to assess application utility. To measure the storytelling effect, a comparative study has been conducted between debate-based learning and storytelling-based learning proposed. The former enabled students to exchange knowledge and have discussions but was still limited to offering documented knowledge accessed through a textbook. In contrast, the latter encouraged students to have interactions with the server system that helps them stimulate the peach tree while continuing to promote user activity in line with the given environment. Both learning approaches were tried on ten teams respectively. Apart from ten teams organized for the storytelling-based learning, there were ten separate teams for the other approach. Both went through a test comprised of 20 questions, and the test results were compared. The questions asked about knowledge acquired from growing the peach tree.

Table 3. Evaluation of Storytelling Effect at App for 'Growing Peach Tree'

Discussion-based Learning				Storytelling-based Learning			
Team	Point	Team	Point	Team	Point	Team	Point
A1	86	A6	80	B1	87	B6	65
A2	87	A7	64	B2	82	B7	88
A3	74	A8	86	B3	81	B8	83
A4	82	A9	79	B4	91	B9	67
A5	87	A10	73	B5	72	B10	88
Average	79.8			Average	80.4		

The average scores for both approaches turned out to be similar with approximately 80 points as shown in Table 3. Even though the average score posted by the group that went through the storytelling-based learning was slightly higher, the difference is not reckoned to be statistically significant. Therefore, the storytelling-based collaborative learning system in the AR, which is suggested by this paper, is reckoned to generate similar learning effects as the existing discussion-based learning.

5   Conclusion

This paper designed and implemented the AR-based collaborative learning system after adding elements of collaboration and storytelling that were not considered much in the existing AR-based learning systems. It suggested and implemented functions of team-making for collaboration, opinion exchanges among members and collaborative growth of content. Furthermore, a rule-based engine for interactive storytelling was developed and introduced to process the storytelling. Also, the open source library ARToolKit was used to implement the AR technologies. As for performance evaluation, learning effects were assessed. The score was 80.4 for learning effects. In general, the scores turned out to be relatively high, which were comparable to the discussion-based learning approach in terms of the learning effect. However, the system proved to be supportive just of the groups who participated voluntarily, leaving teams without voluntary participation with poor results. For further improvements, it is considered to be necessary to introduce a strategic approach to mandate member participation or to allow reward or penalty.

As for the future research direction, additional researches would be conducted to investigate the policy aspect of rewarding team members who make voluntary collaboration and of penalizing those with a low level of participation and to explore system functions, thereby enhancing the learning effect.

Acknowledgement. This research was supported by Basic Science Research Program through the National Research Foundation of Korea (NRF) funded by the Ministry of Education (NRF-2013R1A1A2057943)

References

1. Kaufmann, H.: Collaborative augmented reality in education. In: Proc. of Position Paper for Keynote Speech at Imagina 2003 Conference (2003)
2. Billinghurst, M., Kato, H.: Collaborative augmented reality. Proc. of Communications of the ACM 45(7), 64–70 (2002)
3. Shelton, B.E.: How augmented reality helps students learn dynamic spatial relationships. PhD Thesis, University of Washington, Washington (2003)
4. Billinghurst, M.: Augmented reality in education. New Horizons for Learning (2002)
5. Shelton, B.E., Hedley, N.R.: Using augmented reality for teaching earth-sun relationships to undergraduate geography students. In: Proc. of the 1st IEEE International Augmented Reality Toolkit Workshop, Germany, (2002)
6. Kaufmann, H., Schmalstieg, D.: Mathematics and geometry education with collaborative augmented reality. Computers and Graphics 27(3), 339–345 (2003)
7. Billinghurst, M., Kato, H., Poupyrev, I.: The MagicBook: A Transitional AR Interface. Computers and Graphics 25(5), 745–753 (2001)
8. Billinghurst, M., Kato, H., Poupyrev, I.: The magicbook-moving seamlessly between reality and virtuality. IEEE Computer Graphics and Applications 21(3), 6–8 (2001)
9. Sejin, O., Woontack, W.: Augmented Gardening System with Personalized Pedagogical Agents. In: Proc. of International Symposium on Ubiquitous VR (2007)
10. Roussos, M., Johnson, A., Leigh, J., Vasilakis, C., Barnes, C., Moher, T.: NICE: Combining constructionism, narrative, and collaboration in a virtual learning environment. ACM SIGGRAPH Computer Graphics 31(3), 62–63 (1997)
11. Braun, N.: Storytelling in collaborative augmented reality environments. In: Proc. of International Conferences in Central Europe on Computer Graphics (2003)
12. Diermyre, C., Blakesley, C.: Story-based teaching and learning: Practices and technologies. In: Proc. of 25th Annual Conference on Distance Teaching & Learning (2009)
13. Mello, R.: The power of storytelling: How oral narrative influences children's relationships in classrooms. Journal of Education and the Arts 2(1), 44–65 (2001)
14. Schiro, M.: Oral storytelling & teaching mathematics: Pedagogical and multicultural perspectives. Sage (2004)
15. Koki, S.: Storytelling: The Heart and Soul of Education. PREL Briefing Paper (1998)
16. Norman, D.A.: Things that makes us smart: Defending human attributes in the age of the machine. Basic Books (1993)
17. Jung, J.J., You, E., Park, S.-B.: Emotion-based character clustering for managing story-based contents: a cinemetric analysis. Multimedia Tools and Applications 65(1), 29–45 (2013)
18. Schmidt, V.: 45 Master Characters. Writer's Digest Books (2007)
19. Buttler, T., Lukosch, S.: Exploring the Use of Stories in Patterns of Collaboration. In: Proc. of 45th Hawaii International Conference on System Sciences, pp. 392–401 (2012)
20. Souza, M., Carvalho, D.D.B., Barthy, P., Ramos, J.V., Comunello, E., Wangenheim, A.: Using acceleration data from smartphones to interact with 3D medical data. In: Proc. 23rd SIBGRAPI - Conference on Graphics, Patterns and Images, pp. 339–345 (2010)

Contemporary Architecture Theory and Game Storytelling

Hea-Jeong Jeon and Seung-Bo Park

[1] Mediafish, Seoul, South Korea
[2] Institute of Media Content, Dankook University, Gyeonggi, South Korea

Abstract. The space in the modern age was considered as a fixed space that can be clearly recognized, so we thought that it can be understood at once by separating the subject and object. Thus architects had watched and designed their architecture in a high angle view. However, actually people cannot see them in such views, but they have to understand them by moving themselves. Therefore, the contemporary architects have admitted that the space that gives experience to people would be consequently the montage of scene viewed by them. They cannot be understood in the only one view, but can be only understood by listing the sequence of views seen by people. Hence the most important things would be the position and view direction of the human.

The concept of space that the human understands by seeing and summarizing in the time order is called *scene*. This means that the space and the human body can interact each other and the boundary between them is ambiguous. Thus the space can be considered as the extension of the human body. Digital media, especially video games which is specialized at the spatial storytelling, can also be understood by collecting the scene played by people, and extend the human body. In this paper, we attempt to utilize the concept of contemporary architecture in order to criticize the storytelling of video games in the aesthetical and theoretical aspects.

Keywords: Game, Storytelling, Contemporary Architecture, Digital Media.

1 Introduction

1.1 Motivation

In the current decade, the storytelling of video games has been regarded as the background stories or quests in the games. There have been still some people that have negative thought about video games, and researchers have not been active for accepting video games as aesthetics. To overcome such a situation, analysis by using theoretical tools from other academic fields would be helpful for people to view video games in the other way. Like fiction and film, if there are many tools for criticism about video games, trivially, various reviews would be presented. It would be helpful for people to positively improve their thought and recognize video games as fine arts. Furthermore, it can lead game designers

© Springer International Publishing Switzerland 2015
D. Camacho et al. (eds.), *New Trends in Computational Collective Intelligence,*
Studies in Computational Intelligence 572, DOI: 10.1007/978-3-319-10774-5_14

to try aesthetic experiment in video games, and this would eventually have an positive effect on advance of the field of aesthetic video game.

1.2   Objective

In this paper, we argue that the theory of contemporary architecture can be utilized as a theoretical tool for game storytelling criticism. First we briefly reviewed spatiality of digital media. Then we explain that such a spatiality can be analyzed by considering theories of contemporary architecture, because the spatiality of digital media is clearly generated from participation of users. Finally, we prove that real-time rendering which is an important factor of all games provides 'scene' as concept of contemporary architecture.

2   Background

2.1   Concept of Space in Contemporary Architecture

In the modern age, people thought that space can be separate from experienced person and would not be changed regardless of experience of person. An architecture was located in a fixed position of the space, and the view aspect of the person would be changed with respect to his or her position. After the modern age, however, the world explained by using determinism has dissolved, and the opinions that are analysis of the relative space that a person experiences have become the main stream. Following such opinions, the space that a person does not experience is identical to the space that does not exist. In other words, they argue that we should recognize the space as a variable and relative thing.

In the theories of contemporary architecture, the text in architecture and the participation of human bodies have become important concepts. Wolfgang Iser said that a literary work eventually can be complete after the reading behavior of readers [4]. Likewise, we can consider that a building as a text would also be complete after 'reading behaviours' of person. Gil explained it by quoting Cha's words: the experience of creative works is acheived through the interaction between the text and readers in their consiousness [4].

Gil said that architecture as text provides possibility to the participation of human bodies. He explained that the structure has been *read* and accepted through human bodies. Frankl and Rassmussen also said that the spatial image generated by relationship between a structural space and movement of a human body is an important tool for analysis of an achitectural space. Understanding the achitectural space can be considered as a process that a person summarizes a sequence of images which is captured during exploration inside and outside of the structure. Furthermore, a complicated trajectory generated by the movement of human in three dimensional space can be experienced through the human body with time going. Therefore, we can catch that this aspect would be deeply related to the 'intention to humanize human' [4].

Such a sequence of images experienced by person through his or her body is called *scene*. In order words, the scene is a space that consists of a collection of views. This concept definitely denies the aspect of modern achitecture, which argues that a person can be an independent subject of observation and can understand a structure in a high angle view. As we mentioned above, a complicated trajectory generated by the movement of human can be treated as the space itself, and this space can be considered as the scene constructed by experience of a person.

2.2   Spatiality of Digital Media

The concept of space used in the contemporary architecture field is similar to the property of digital media. Especially, we can induce it by reviewing various studies about digital virtual worlds.

In according to Murray, digital medium is non-linear, which means that it does not follow chronological order unlike books and films, and has spatiality in itself. She said "The new digital environments are characterized by their power to represent navigable space. ··· It is in fact independent of the computer's ability to display maps, pictures, or even three-dimensional models." [1]

The virtual world is the representative space that can be generated from digital media. Manovich defined the virtual world as follows: "By virtual worlds I mean 3D computer-generated interactive environments. This definition fits a whole range of 3D computer environments already in exsitence: high-end VR works which feature HMD and photo realistic graphics; arcade, CD-ROM and on-line multi-player computer games; QuickTime VR movies; etc." [2]

Seel said that the virtual world can be meaningful only if a body is related to it. He employed the concept of cyberspace in the narrow sense of a mechanically produced spatial state that changes its vistas in coordination with the bodily movement of beholders [3].

Of course, we have to use our bodies in order to experience real world. However, our existance is not strictly necessary for the existance of the real world. In a point of modern view, the space can exist, even though we do not experience it and do not know about it. On the other hand, in a point of contemporary view, the space can be meaningful only if the participation of the body happens. In order words, there would be no space if there were no movement, so the space can be only recognized by the movement. This concept of user-oriented space leads human-oriented design in current architeture.

Furthermore, the space of virtual worlds definitely agrees the concept in contemporary architecture. It cannot be presented if the user does not perform any action. All in the virtual world stores by using binary code without any physical shape, and they would be shown in the user's view through real-time rendering while he or she interacts with them. In order words, the space in the virtual world is generated and exists temporarily and interactively by the participation of the user.

The subject in the virtual world is not also considered as 'fixed user'. Since the user interact with the virtual world, he or she only becomes a position of the virtual camera in order to show a rendered scene in real-time. The fact that the user's intention changes the space of the virtual world is clear, but the user also become a virtual point selected in that space. Such a close interaction that the subject and object cannot be separable and only the actions between them can be exchanged would become stronger in the virtual world than in the previous digital media.

This means that the behavior exploring the space regulates the user and extends one's experience and body.

2.3   Spatial Storytelling of Video Games

Video games have a clear characteristic that the user is actually involved in the context and the progress through the active participation and control. The previous texts were also considered that they can interact with readers. However, while the previous texts are presented by physical objects including published media or structures so they have a limitation for interaction with the user, video games can destroy such a limitation.

At this point, Jeon mentioned that the text as a physical object is a finite object that is completed by the author so that it is limited spatio-temporally and self-contained [8]. On the other hand, video games are flexible spatially and temporally, which is clearly different from the others. In the video game, the boundaries abong authors, audiences, actors, and characters become ambiguous because of its interactive property. Furthermore, in the case of massive multi-player online role playing game (i.e. MMORPG), its narratives depend on the collective experiences of the users [8].

A video game is one of the most flexible text in the presented texts. Especially, the online video game that thousands of users concurrently plays through internet such as MMORPG is a common but successful one of virtual worlds.

Therefore, storytelling of the video game cannot be intended by its creators, but can be directly produced by the users that explores the world in the game. The remained question is how to design such a world in the author's aspect. The narrative intended by the authors can be constructed by using the background story, system setting, and world campaign setting, and this would work as a guideline for exploration of the users.

This setting of the game world is more important in the game that has a huge and complex virtual world such as MMORPG that the others, because an elaborate designed setting for the game world should be related to all components in the game. By premising this, the user can understand the spaces collected from his or her exploration as the whole world, and this leads the user to be immersed in the game. Moreover, the users can use the setting as a guideline, and they can have a proper experience of the story during the exploration of the game world.

3   Real-Time Rendering and Scene

Like the other virtual worlds, video games are also presented by using real-time rendering. Real-time rendering is one of the interactive areas of computer graphics, it means creating synthetic images fast enough on the computer so that the viewer can interact with a virtual environment.

Fig. 1. The church designed by Ando Dadao

This real-time rendering necesserily produces the space for the scene. As we mentioned above, the scene means the collection of images through the body movement. Several architects pursue scenes in their contemporary architectural work.

Fig. 1 shows the church designed by Dadao. People should pass a narrow and crooked concrete road to enter the church. At the end of the concrete road, people would see the cross-shaped light in the center of church. This is a good example that the scene in the structure provides a well-organized narrative to the peoples.

The video game also provides the space for the scene, because it provides the specific image of the space at the moment of interaction through the real-time rendering system. Therefore, the users should understand the space by moving their avatar or virtual camera. Furthermore, the video games can be considered as spatial storytelling contents which put their story in the space. Thus, the exploration of the space in the game world would be identical to the story experience.

4   Conclusion

Several games provides a virtual camera that can be unlimitedly controlled by users, so that they can see the game world in a high angle. However, if there is an avatar which represents the user in the game, the camera would be also limited to the avatar. Hence the game should have a capability to deliver its story by only using the scenes without a high angle shot. Especially, many games that consist of the only one way to proceed (i.e. interactive drama, e.g. The Last of Us) are designed by using the space for the scene.

In short, the real-time rendering in the game provides scenes, and they becomes the main characteristic of the spatial storytelling of the game. Therefore, when we want to analyze the storytelling of the game, we should analyze the way how the scene delivers the story and let the user understand the space.

Acknowledgement. This research was supported by Basic Science Research Program through the National Research Foundation of Korea (NRF) funded by the Ministry of Education (NRF-2013R1A1A2061737).

References

1. Murray, J.H.: Hamlet on the Holodeck: The Future of Narrative in Cyberspace. The Free Press, New York (1997)
2. Manovich, L.: The Language of New Media. The MIT Press, Massachusetts (2002)
3. Seel, M., Farrell, J.: Aesthetics of Appearing. Stanford University Press, California (2005)
4. Gil, S.H.: The Aesthetics of Receptiuon and Contemporary Architecture. Spacetime publishing, Seoul (2003)
5. Lee, I.H., Ko, W., Jeon, B.K., Kang, S.H., Jeon, K.R., Bae, J.Y., Han, H.W., Lee, J.Y.: Digital Storytelling. Golden Bough Co., Seoul (2003)
6. Lee, W.G.: Digitalized Image and Virtual Space. Yonsei University Press, Seoul (2004)
7. Jeong, S.Y.: A Study on Spatial Storytelling of the Educational Virtual World, Thesis of Master Degree, Ewha Womans University, Seoul (2010)
8. Jeon, G.R.: A Study on the Actualization of Interactive Text: Analysis of the Massively Multiplayer Online Role Playing Game Playing. Korean Journal of Journalism & Communication Studies 48(5), 188–213 (2004)

Calibration of Rotating 2D Laser Range Finder Using Circular Path on Plane Constraints

Laksono Kurnianggoro[1], Van-Dung Hoang[1], and Kang-Hyun Jo[2]

[1] Graduate School of Electrical Engineering, University of Ulsan, South Korea
{laksono,hvzung}@islab.ulsan.ac.kr
[2] School of Electrical Engineering, University of Ulsan, South Korea
acejo@ulsan.ac.kr

Abstract. Rotating a 2D laser range finder is a common way to obtain the 3D data of a scene. In the real system, the rotation axis is not always coincide with the sensor's axis. Due to this matter, it is important to know the rotation axis of a rotating 2D laser range finder system to obtains an accurate 3D measurement. In this paper, a method for determining the rotational axis and its relation to the sensor's position is presented. The rotational axis is approximated by analyzing the surface normal of several sensor's positions with different rotation angle. The sensor's positions are projected to the rotational plane to obtain a circular path. The circular path is approximated using a circular fitting method. A comprehensive analysis from a simulation system is presented in this paper to show the performance of the proposed calibration method.

Keywords: Laser range finder, extrinsic calibration, circular fitting.

1 Introduction

Environment understanding problem is an important task for autonomous systems such as mobile robot or unmanned vehicle. One of the important sensing devices for environment understanding is 3D scanner. This kind of device provides geometrical information which is useful for many tasks including object recognition, obstacle avoidance, path planning and localization.

There are many kinds of 3D measurement device. Stereo camera system is widely used as 3D measurement device, however it performance is suffered from noises and only works well on a limited distance. Another affordable 3D scanner is time-of-flight camera and structured light camera which is commercially available. These two kinds of 3D scanners are not suitable for long range measurement due to the characteristic of the sensor. Velodyne HDL is a commercially available high quality 3D laser range finder. Due to the high cost, it is not widely used.

A 2D laser range finder (LRF) system is widely used. It has high precision and long range scanning distance. However, the scanned 3D data from 2D LRF sensor is limited only in one line scan. To obtain a true 3D scanning result, rotating the 2D LRF is an economical solution.

© Springer International Publishing Switzerland 2015
D. Camacho et al. (eds.), *New Trends in Computational Collective Intelligence*,
Studies in Computational Intelligence 572, DOI: 10.1007/978-3-319-10774-5_15

In this paper, a method for determining the rotation axis and rotation center of a rotating 2D LRF is presented. In the rotating LRF system, the sensor's origin travels a circular path that lies on an oriented plane. The plane's orientation determines the direction of the rotational axis. Meanwhile the rotational center is determined by the center point of the circular path that lies on the rotational plane.

The calibration procedure is consists of three steps. The first step is determining the sensor's position at several different rotation angles. The second step is determining the parameter of the rotational plane. This can be done by employing a surface normal estimation method. The last step is determining the center of the rotation system. Knowing the plane's parameter, the circular path can be determined by projecting all the sensor's positions on the rotational plane and then performs a circular fitting method over those projected points.

The paper is organized as follows. In the next section, several related works are explained. A brief explanation about the proposed method is covered in the section 3. In section 4 the simulation setup and its results are presented. Finally, the summary of the proposed method and the contribution of this paper are presented in section 5.

2   3D Scanner Using Rotating LRF System

Calibration method of rotating 2D LRF system has been studied intensively in past few years. A calibration method using special 3D pattern is presented in [1]. In that paper, the calibration pattern is a stereoscopic checkerboard. It is a 3D pattern with several black cubes positioned in a checkerboard formation. The calibration system consists of a rotating LRF and a camera. To do the calibration, corresponding features of intensity images and laser scans are extracted and then analyzed to determine the calibration parameters.

Another form of special calibration pattern is used in [2]. A calibration board with right angle outline is used. The shape of the calibration board several observable properties from the scanned 3D points. The important properties are the linearity property, the perpendicular property and the co-planarity property. The calibration parametersare calculated using those observed properties.

There are exists more calibration method using special 3D structure such as in [3]. However using such kind of calibration pattern is practical. It is not an easy task to provide the 3D calibration structure with accurate shape.

A calibration method using conventional checkerboard pattern is presented in [4]. The calibration setup needs a camera to compute the relative pose of the LRF at two different rotated angles. The LRF poses are approximated by performing camera-LRF calibration method using point-plane constraints at each rotation angle. Having two poses of a rotated LRF, the rotation axis is extracted using screw decomposition method.

3   Calibration Procedure

A rotating LRF system can be visualized as in Fig.1 (a). In the ideal case, the LRF sensor travels along a circular path on a rotational plane. In general, the rotational axis may not coincide to the axis of LRF sensor. Another problem is that the origin of a LRF sensor may not located in the same position with the center of the rotational path. The aim of this work is to determine the rotational axis ($\bar{\omega}$) and the rotation center (t) of the rotating LRF system.

3.1   LRF Pose Extraction

The first step of calibration procedure is obtaining several positions of a LRF sensor on the rotating system. This procedure can be done by calibrating the LRF sensor to a camera. For each rotation angles, the calibration method is performed to define the relation between the LRF pose and the camera pose. Hence for n-rotation angles, n-poses of the LRF system relative to the camera's coordinate system are obtained.

Calibration of LRF system to a camera system can be done using one of the methods presented in [4,5,6,7,8]. The method presented in [5] is the most commonly-used solution for this kind of task. However, an experiment in [6] shows a novel method that gives a more efficient solution compared to the method presented in [5].

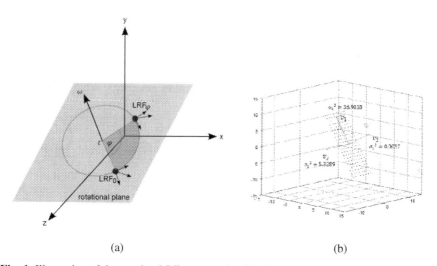

(a)   (b)

Fig. 1. Illustration of the rotating LRF system, the circular rotational path lies on the rotational plane (a). Eigen-vector which corresponds to the smallest Eigen-value represents the surface normal (b).

3.2   Rotational Plane Model

The second step of the calibration procedure is estimating the rotational plane model. This plane model gives the orientation of the rotational plane.The plane model represents the direction of the rotational axis as shown in Fig.1 (a). Having several LRF poses from several rotation schemes, the plane model can be computed using the surface normal computation method.

A lot of surface normal computation methods are available nowadays. A simple computation method is shown in [9]. However, this method is not robust against noise which may affect the final result of the calibration method. Several optimization methods which are robust against noise are presented in [10].

Having a set of LRF positions P, with $P = \{L_1, L_2, ..., L_N\}; L \in \mathbb{R}^3$, a covariance matrix M can be formulated as shown in (1).

$$M = \begin{bmatrix} cov(P_x, P_x) & cov(P_x, P_y) & cov(P_x, P_z) \\ cov(P_y, P_x) & cov(P_y, P_y) & cov(P_y, P_z) \\ cov(P_z, P_x) & cov(P_y, P_z) & cov(P_z, P_z) \end{bmatrix} \tag{1}$$

where $cov(P_x, P_y) = \frac{1}{N}\sum_{i=1}^{N}(P_{x,i} - \bar{P}_x)(P_{y,i} - \bar{P}_y)$.

The covariance matrix consists of covariance values from each combination of x,y and z element of the set of points P. Having the covariance matrix, the surface normal of the rotational plane can be obtained using the singular value decomposition (SVD) method or the Eigen-value decomposition method. Both of the methodsgive similar values.

The SVD method produces U, S, and V^T matrices. The singular values are stored in the diagonal matrix$S$. Each value in the matrix S corresponds to a vector in each column of the left-singular vector U. In the concept of principal component analysis (PCA) method, vectors in the matrix U represent a new coordinate frame. In this new coordinate frame, the data are spreading with variance in each new axismentioned in the matrix S. The illustration of this concept is shown in Fig.1 (b).

$$\hat{\omega} = U_{ki^*}; k \in \{1,2,3\} \tag{2}$$

$$i^* = \underset{i}{\arg\min} S_{ij} ; \text{ where } i = j , \ i \in \{1,2,3\}$$

The direction of the rotational axis $\hat{\omega}$ can be formulated as (2). Having the values of$U$ and S from the decomposition of covariance matrix M, where $USV^T = M$, the direction of the rotational axis is the column vector that corresponds to the minimum value stored in the diagonal of the singular values S.

3.3   Estimation of the Rotational Path

The trajectory of the LRF origin in the rotating system is a circular path. Having the set of points of the LRF position from different rotating angles, the circular path that lies on a rotational plane can be estimated using a circular fitting method. Several circular fitting methods and their performance are examined in [11]. In this work, the circular fitting method follows the method presented by Bullock[1].

The circular fitting method works in the two dimensional space. The obtained LRF poses should be projected to the rotational plane before they are used in the circular fitting method.   Having a set of LRF positions P, its projection on the rotational

---

[1] Least square circle fit,
http://www.dtcenter.org/met/users/docs/write_ups/circle_fit.pdf

plane can be formulated as P_{PCA} in (3). The projected points P_{PCA} are computed by multiplying the matrix V^T from the SVD result to the zero centered LRF's position points.

$$P_{PCA} = V^T(P - \bar{P}) \tag{3}$$

The resulting values in (3) is originally a set of points in $\mathbb{R}^3$. However since the components in the axis that represents the plane's surface normal are always zero, these resulting points can be treated as set of points in $\mathbb{R}^2$. The component of P_{PCA} in the first and the second axis are represented as X and Y respectively. The center of the rotational path is defined as (4), where the values of u_c and v_c is obtained by solving (5). The zero centered values of X are represented as $u = X - \bar{X}$ while the zero centered values of Y are represented as $v = Y - \bar{Y}$.

$$R_c = [u_c, v_c]^T + [\bar{X}, \bar{Y}]^T \tag{4}$$

$$\begin{bmatrix} \Sigma_{uu} & \Sigma_{uv} \\ \Sigma_{uv} & \Sigma_{vv} \end{bmatrix} \begin{bmatrix} u_c \\ v_c \end{bmatrix} = \frac{1}{2} \begin{bmatrix} \Sigma_{uuu} + \Sigma_{uvv} \\ \Sigma_{vvv} + \Sigma_{vuu} \end{bmatrix} \tag{5}$$

where $\Sigma_{uv} = \sum_{i=1}^{N} u_i + v_i$.

Using this circular fitting method, the radius of the circular path r can be formulated as defined in (6).

$$r^2 = u_c{}^2 + v_c{}^2 + \frac{\Sigma_{uu} + \Sigma_{vv}}{N} \tag{6}$$

The rotational center position estimated by the circular fitting method is a position in the PCA coordinate frame. In order to convert this point to the global coordinate frame, a conversion from R_c to t is defined as (7).

$$t = V[R_c{}^T\ 0] \tag{7}$$

4   Simulation and Results

The proposed calibration method was tested on a simulation environment using MATLAB software. In the simulation, a set of scanned points lie on a same plane was generated as a groundtruth data. In the ground truth data, the points were generated by changing the radius of the rotating system and adding noises to the LRF's positions for each simulation case.

For simplicity, the simulation scenario assumes that the direction of the rotational axis is parallel to the y-axis of the LRF's coordinate frame. The illustration of this scenario is shown in Fig.2. Hence the ground truth points are defined as (8). The rotation angle on the rotating platform is denoted as φ while the measurement angle in a LRF's pose is denoted as θ.

$$^{GT}P_{\varphi,\theta} = \left[(z + r)\tan\varphi \quad \frac{z + r - r\cos\varphi}{\cos\varphi}\tan\theta \quad z\right] \tag{8}$$

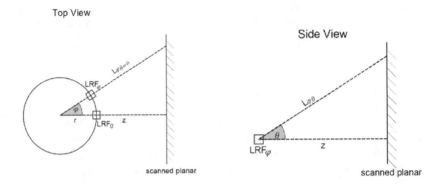

Fig. 2. Illustration of the simulation's scenario

To simulate the data given by the LRF sensor, a set of measurement data are generated according to the ground truth points. The simulated range data are generated according to (9).

$$L_{\varphi,\theta} = \frac{z+r-r\cos\varphi}{\cos\varphi\cos\theta} \tag{9}$$

In the simulation, the effect of several combinations from rotating system's radius and the LRF's location noise are examined. In this work, several metrics are used to define the quality of the proposed method. These metrics are point to plane distance error, root mean square (RMS) error, point to plane distance error'variance, angular error, and radius error.

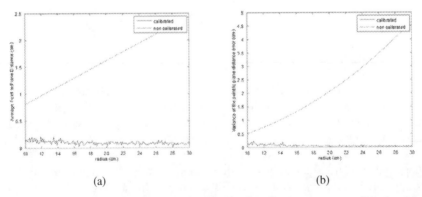

(a)   (b)

Fig. 3. In the un-calibrated rotating system, the point to plane distance error grows linearly with the radius (a). The performance of the calibrated system is not affected by the radius variations (b).

Table 1. Analysis of the error in the calibrated rotating system by varying the radius

Metric	min	max	mean	median	mode	σ^2
RMS	0.1581	1.301	0.4941	0.4598	0.1518	0.2014
P2P error	0.02772	0.2152	0.09729	0.09223	0.02772	0.03385
Angular error	0.004394	0.8935	0.288	0.2539	0.004394	0.181
Radius error	0.7591	2.204	1.403	1.384	0.7591	0.2911

In the first scheme of the simulation, several radiuses with a fixed LRF's location noise are used. The LRF's location noise is set as a random Gaussian noise with variance 0.1. As shown in Fig. 3 (a), the point to plane distance error for the proposed method is relatively constant. In contrast with the calibrated system, the non-calibrated system produces higher point to plane distance error as the radius getting bigger. Although the point to plane distance error for the non-calibrated system is relatively low, the non-calibrated system is not reliable whenever the radius is big. Fig. 3(b) shows that as the radius getting bigger, the variance of the point to plane distance error is getting bigger. This phenomenon is occurred since in the non-calibrated system the point to plane distance error is low for small rotation angle. However it is become very high whenever the rotation angle is big. More detailed results are shown in the Table 1.

Table 2. Analysis of the error in the calibrated rotating system by varying the noise

Metric	min	max	mean	median	mode	σ^2
RMS	1.227	14560	1625	695	1.227	2300
P2P error	0.04584	1.019	0.4647	0.4677	0.04584	0.2034
Angular error	0.04549	3.06	0.8458	0.6963	0.04549	0.6291
Radius error	1.143	9.371	6.051	6.469	1.143	2.073

The second scheme of the simulation is using several noises value for generating the LRF's position whenever the radius of the rotating system is maintained to be constant. The radius error is getting bigger as the noises getting larger as shown in Fig.4 (a). The radius error is the different between the computed radius and its ground truth. This is occurred since the noises affecting the circular fitting result. Whenever the noise is big, the input for the circular fitting are scattered points. Those scattered points are not forming a circular path. This is the reason why the radius error is getting bigger as the noise getting bigger. Table 2 shows similar result for the angular error.

In the next simulation scheme, the effect of the noise on the rotating system's radius is examined. Using several combinations of noise and radius value, Fig.4 (b) shows that the system delivers a good result whenever the variance of the noise is below 5% of the rotating system's radius, which will gives the maximum RMS error around 8 cm.It is means that whenever the radius of the rotation system is 10cm, the estimation of the camera to LRF's position should not more than 5mm.

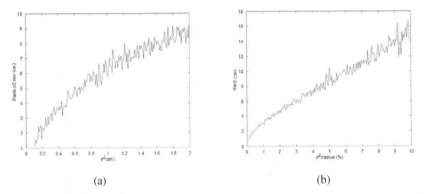

(a)   (b)

Fig. 4. Error in the radius estimation is getting higher as the noise getting higher (a). The RMS grows as the ratio between noise's variance to rotating radius getting bigger (b).

5   Conclusion

A novel calibration method for rotating LMS system is presented in this paper. The calibration method is consists of three steps, approximating the LRF poses for each rotation angle, determining the rotational plane model, and approximating the rotational path. The quality of the proposed method is tested on a simulation environment under several cases. The simulation shows that the quality of the calibration method is strictly related to the presence of the error in the estimated LRFpose. The proposed calibration method requires a high precision of LRF pose estimation method. Due to this matter, for the future work, it is important to develop a method that can estimate the LRF's position with a high precision by eliminating the camera from the calibration system.

References

1. Jia, T., Fu, M., Yang, Y., Zhu, H.: Calibration of a 3D laser range finder and a camera based on stereoscopic checkerboard. In: International Conference on Vehicular Electronics and Safety (ICVES), Beijing, pp. 92–96 (2011)
2. Fan, H., Li, G., Dong, L.: The calibration algorithm between 2D laser range finder and platform. In: International Joint Conference on Computational Sciences and Optimization (CSO), Hainan, pp. 781–783 (2009)
3. Yang, G., Zhengchun, D., Zhenqiang, Y.: Calibration method of three dimensional (3D) laser measurement system based on projective transformation. In: International Conference on Measuring Technology and Mechatronics Automation (ICMTMA), pp. 666–671 (2011)
4. So, E., Basso, F., Menegatti, E.: Calibration of a rotating 2D laser range finder using point-plane constraints. Journal of Automation, Mobile Robotics & Intelligent Systems 7, 30–38 (2013)
5. Zhang, Q., Pless, R.: Extrinsic calibration of a camera and laser range finder (improves camera calibration). In: International Conference on Intelligent Robots and Systems (IROS), pp. 2301–2306 (2014)

 6. Vasconcelos, F., Barreto, J.P., Nunes, U.: A minimal solution for the extrinsic calibration of a camera and a laser-range finder. Pattern Analysis and Machine Intelligence (PAMI) 34, 2097–2107 (2012)
 7. So, E., Menegatti, E.: A unified approach to extrinsic calibration between a camera and a laser range finder using point-plane constraints. In: Proceedings of the 1st International Workshop on Perception for Mobile Robots Autonomy (2012)
 8. Hoang, V.-D., Hernández, D.C., Jo, K.-H.: Intelligent Information and Database Systems. In: Intelligent Information and Database Systems, pp. 561–570. Springer International Publishing (2014)
 9. Kurnianggoro, L., Jo, K.H.: Free Road Space Estimation Based on Surface Normal Analysis in Organized Point Cloud. In: International Conference on Industrial Technology (ICIT), pp. 609 – 613 (2014)
10. Klasing, K., Althoff, D., Wollherr, D., Buss, M.: Comparison of surface normal estimation methods for range sensing applications. In: International Conference on Robotics and Automation (ICRA), pp. 3206–3211 (2009)
11. Umbach, D., Jones, K.N.: A Few Methods for Fitting Circles to Data. IEEE Transactions on Instrumentation and Measurement 52, 1881–1885 (2003)

A Case Study of Database and Information System on Mobile Application for Tourism

Pongpuk Theppawong and Metawat Kavilkrue[*]

Faculty of Engineering, North-Chiang Mai University, Chiang Mai, Thailand
asagura_boy_pongpuk@hotmail.co.th,
metawat@northcm.ac.th

Abstract. Internet is more prevalent today, especially on mobile phone. The internet can be used anywhere to make it easy to find the information at any time. At present, the number of tourists has increased, especially for travelers who like to travel in Thailand. The problems are that many of the tourist attractions, travel search, just a passport. The map is not enough for the journey of a group of tourists. It has solved the problem by using the convenience of the Internet to help in identifying the attractions in the vicinity. Paths where tourists want to go with a GIS such as restaurants, hotels, etc. By the Sea Island Resort National Park displays the results based on the needs of the tourists on the web internet.

Keywords: Web Base, Tourist Attraction, Mobile Phone, GIS.

1 Introduction

At present, the Internet is more prevalent in society with the technology of mobile phone. It makes almost all phone models can use the Internet as advertising. It can be shared all around the world easily. Whether advertising sites selling products, we do all the time. Therefore, tourists are attracted to destinations through the Internet, in particular Thailand. Which has increased as many tourists as Thailand is known as the city of temples, notably struck with foreigners very much as a result, more and more tourists to Thailand. And hence the problem of wandering is not to stay. Where to eat all visitors have the original passport with a tourist map, but do not know where they are now and that place is also close to other places. The area this has led some of these problems, analyze them and find solutions to facilitate the tourists as possible. That is, making a Web base storage landmarks. Of tourists, such as hotels, resorts, restaurants and other tourist attractions tourists want to go. By Say What Travelers want to go and places associated with that location in the area around the web base on the Internet with the use of GIS as an aid in the said location.

---

[*] Corresponding author.

© Springer International Publishing Switzerland 2015
D. Camacho et al. (eds.), *New Trends in Computational Collective Intelligence*,
Studies in Computational Intelligence 572, DOI: 10.1007/978-3-319-10774-5_16

165

2   Problem Definition

From Problems of Tourist Groups in Thailand. Can be classified as follows

1. Tourists are not aware of other areas, then there's any place else in the case of a traveler needs to stay full because tourists are traveling as a tourist attraction. Or the source of the book and is highly recommended that the property room is full before. Other travelers to a later.
2. Tourists do not know Attractions that are also attractive in the immediate vicinity of the spot where tourists want to go. Besides famous attractions then. It still has the most beautiful sights as the main tourist attractions which is still not as popular or famous.
3. Tourists do not know where they are on the map many tourists who stray. Because the map may be easier to understand the need to go way north, but the fact that the city is home to several buildings layer visibility. I looked and routes and stalking noticed in traveling to a place which travelers need to.
4. It has no program or website that can answer the locations of a people. Because many citizenship of each person, but many are interested in making the place. Travelers to individual needs vary.

3   Current Uses of GIS in the Tourism

There are two categories for the use of GIS system in tourism, public use and management use. The public wants to find geographic information about a place before they go there. They wants to know where things are located, what amenities are available, what the climate is like, and be able to do site specific searches to find information. This research was to research a case study to analyze the problems. Management side may be done by individual operators, a tourism group, or by the local municipality. Management users want to query the system for where customers are coming from.

Group of tourists in northern Thailand are sampling various needs. Traveler's needs and problems encountered on a regular basis until a bad attitude in tourism in Thailand. The sampling was carried over from tourists in northern areas is key, and with this in order to meet the needs of the tourists, most likely it has been studied in many ways. Aspects be compiled and used to assist in the analysis. Meet the needs of tourists. Extras by information that meets the needs of the people on this show, Web base on the Internet, which make it easy to locate more. It also shows the locations in the vicinity of the location where the tourists are.

3.1   Study Needs of Tourists in Northern Areas

This study of the group of tourists in Thailand with tourism interests in the region. This was a relatively large number of inquiries and find out the requirements for each aspect, such as accommodation, restaurants, etc., which in most northern climates it is not hot as in other regions. This makes a lot of foreigners and hence also meet tourists face problems in various fields that whether it's the journey accommodation in restaurants, etc. This is the main factor of tourists will come to Thailand and to meet the needs of travelers. Conduction problem and needs to analyze and respond to tourists can be identified as the following table.

Table 1. Requirements of the tourists in Northern of Thailand

Requirement of tourists	Management Tourists Requirement
In this area there are lots of other tourist attractions?	The tourist attractions that tourists want to be tourist look?
In this area, what are their restaurants?	Restaurant tourists the restaurant is a star? At any given time? Small restaurant or big restaurant?
In this area has hotel that you want to stay?	The place that tourists want to stay a few levels where famed stars look like? A resort or something? And want to stay in the area where such a nature in the city or outside the city?
In this area has massage or spa?	Tourists need a massage service as well as Thai plan? Traditional? Visitors to the spa features?
Going to places where there is a festival?	Tourists to the festival like? Festivals, need long term or short term? Want festival held every week, every month or every year.
Tourists want to travel that way?	Manually travel or need transportation? In the case to a distant location. And the difficulty (Which corresponds to the path shown in Web travel).

From the table to see the demand characteristics of the groups of countries in northern Thailand, these travelers want? Places which meet the requirements in this area some of which were just a short time travel. Some groups also travel a long period these data correspond to a different residence. Travelers looking to stay and to analyze the needs of tourists in each category are shown in Table 1. It is just a large sample which makes the overall needs of the tourists. In order to contribute to the next stage or details the characteristics of the group to the category. The column 2 of Table 2 that show details will be taken to meet the needs of individual travelers as well.

3.2   Study Northern Areas

Because in the northern region consists mainly of mountains which is rich in natural? The various species of trees and dogmatic with an environment that it dries and not too cold. It makes plants trees grow well and attracting the eyes of many tourists. Especially during the winter

First, in the study area in the North are the study sites significant in each of the North. Each province will also be featured in the attraction in its own format. Further study of the major attractions were also studied other attractions. The nearby attractions are the source for great value on travel of tourists at a time.

Second is a study of the accommodation, restaurant and other services. It's necessary to tourist groups by studying the key points, such as at the source of every tourist attraction in north point according to the center for nightlife around the town center to cater for tourists who do not like the hustle and bustle of noise. For example, natural infection and various Resorts built by him to provide residents with wholesome atmosphere of the north. As well as exploring the surrounding radius of 1 kilometer. For services it is full. Tourists will have a place to stay the next.

Third is a cultural study of each province which are the important things to follow, whether Because each province has a different culture. Should know it's for a visiting tourist attraction. Shelter is not the tradition of each province.

Fourth is studying the environment and weather. Because these factors affect tourists are sufficiently. Acceptable, because most of the tourists come from countries in the North, where there is cold weather. Not been to hot weather in Thailand. This is so as to accurately reflect the environment for both business and leisure travelers to visit the different places in Northern.

Fifth, studied the routes travel to various destinations to speed travel for both business and leisure travelers and notice the focus during the journey to the places that tourists want. For tourists who want to travel by herself to feel the real tourism.

3.3   Web Based GIS

In this research studied, only the defined area. That is, the northern of Thailand by studying the GIS data to locate the various attractions. Including places has been studied before, above. Indicated on the map, so easy to search for the important position of the tourists and kept in a Web base on the Internet. This easily fined locations in between. The weather is accommodation restaurants and attractions, festivals in the Northern provinces. The data is displayed as a map with our location indicates the tourists at the moment. When visitors want to find, you can find it, and the system will show the places with the position.

- First step, collect all natural, cultural and socioeconomic data concerning the study are to form a database.
- Second, data that regarding to study area are transfer into PC by Scanner and digitizers and these data related with the database which is formed in the first step to constitute a skeleton of a tourism information system.
- Third, obtain photos of the historical and cultural building, open areas and hotels to connect with the graphical data in the tourism information system.
- Fourth, prepare a software; which is devoted to the tourists query and a software that supply to tourist which uses tourism information system that easiest and the fastest by icons and signs.

- Fifth, in this step of the study use required software to transform the tourism information system, which a formed with all literary, visual, graphical data, into a system can supply a query on interest.
- Sixth, to set up this system by relating the software-hardware and client-server.

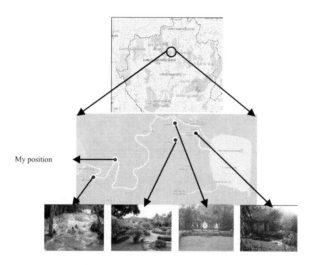

My position

Fig. 1. Tourism on GIS

This figure 1, travelers see to tourist and relate place on geographic which show tourist on web based by mobile phone to search this data.

3.4   Geographic Query Work Flow and System Flow Chart

The integration of multiple technologies, including tourism information system, Internet and Geographic Information System (GIS) has given an exciting new access to tourism information. As an information based society, tourist value systems and services that inform them about the location of touristic objects like hotels, restaurants, theatres, museums, resorts, etc. This helps the tourist to find the most relevant accommodation or to locate the position of a specific tourist place. The existing tourism information systems do not reflect this requirement.

This section is system design of query web base for show important tourist and author place relative to request place of tourist.

This figure 2, its work by input requirement form user. By enter data into web after that system will select module of GIS through to translate. This module will bring data form tourist database and GIS database of northern in Thailand to create tourist list According to their position by saying that the user needs to go through a Web base the user fills out earlier.

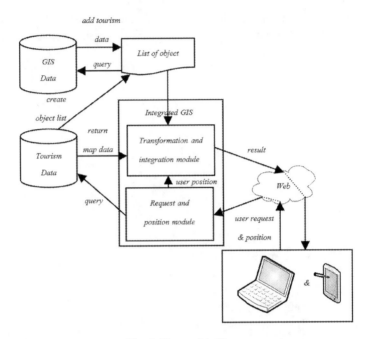

Fig. 2. Geographic Query

4   Experiment/Evaluation

The areas consists 9 provinces in northern of Thailand by climate development authority. The numbers of users or tester an application is 1,000 persons. We use Google maps API to show the relevant places for tourist. The input of this application is a position value (Latitude and Longtitude) of user.

In the experimental results, this has created a place by bringing the needs of tourists. In order to meet the requirements by showing various information. By each of the requirements by the following

Of the above experiment, it turns out that the results of the show to meet the needs of travelers in each category. Able to demonstrate clearly the case for accommodation or traveler's needs and place around a radius of 1 km, which gives visitors a place to meet in case that location . Crowded or full cannot go to such places have. You can also find destinations more easily. It's without having to open the map to find attractions to waste time chasing with the collected information, attractions, and other locations. Critical of the GIS system will always see clearly location and more. Simply connect to the Web base on the Internet, so gathers information to display on this map. The answer to the tourist's very well. With the ease of search that is available on mobile phone wherever you are. That can connect to the Internet.

Table 2. Results based on needs of Travelers

Requirement of tourists	Result
In nearby area attractions Anything else?	Show attractions within a radius of 1 km of tourists. And display locations With respect to or in the same manner.
In nearby area have restaurants in?	Show Restaurant with looks to match the needs of the class. And similarly for tourists. Includes a restaurant within a radius of 1 kilometer from the desired position.
In nearby area offers a level of luxury with the need to stay again?	Show accommodation in a direct manner to the needs of tourists. It shows the level of accommodation that. And properties that are similar to the requirements of users in a radius of one kilometer.
In nearby area with a massage or spa, right?	Show more service based on the needs of the position or space to store and display a surrounding radius of 1 kilometer.
Going to places where there is a festival?	Show attractions or festivals. It tells To organize a festival that The position on the diagram, the geography of each province in the north.
For travelers as it is located at the point on the chart?	Shows the location of that moment on the geographic map.

5   Conclusion

For guidelines on how to develop can lead to the development needs of the target groups of the exhibition possible. The traveling demand need a place that facilitate different. This research also has the problem of facilitating travel for tourists, because just the information displayed on the map are not able to meet the needs of tourists because of some groups may want a private guide the place Interest to them For this reason, the navigation system can be developed as a guide for tourists who want to travel on their own. Whether by foot, bus, designed for those interested in future study.

References

1. Duran, E., Seker, D.Z., Shrestha, M.: Web based information system for tourism resort, Turkey
2. Camacho, D., Borrajo, D., Molina, J.M.: Intelligent travel planning: a multiagent planning system to solve web problems in the e-tourism domain. Autonomous Agents and Multi-Agent Systems 4(4), 387–392
3. Camacho, D., Aler, R., Borrajo, D., Molina, J.M.: Multi-agent plan based information gathering. Applied Intelligence 25(1), 59–71

4. Jung, J.J.: Exploiting Multi-Agent Platform for Indirect Alignment between Multilingual Ontologies: a Case Study on Tourism Business. Expert Systems with Applications 38(5), 5774–5780 (2011)
5. Jung, J.J.: Ubiquitous Conference Management System for Mobile Recommendation Services Based on Mobilizing Social Networks: a Case Study of u-Conference. Expert Systems with Applications 38(10), 12786–12790 (2011)
6. Camacho, D., Molina, J.M., Borrajo, D.: A multiagent approach for electronic travel planning. In: Second International Bi-Conference Workshop on Agent-oriented Information Autonomous Agents and Multi-Agent Systems, vol. 4, pp. 387–392 (2001)
7. Camacho, D., Borrajo, D., Molina, J.M., Aler, R.: Flexible integration of planning and information gathering. In: International Symposium on Methodologies for Intelligent Systems, ISMIS 2002 (2002)
8. Camacho, D., Borrajo, D., Molina, J.M.: TravelPlan: A multiagent system to solve Web electronic travel problems. In: Workshop on Agent-Based Recommender Systems, Fourth International Conference, Autonomous Agents and Multi-Agent Systems, vol. 4(4), pp. 387–392 (December 2001)

A Novel and Simple Management System of Interactive System Using Google Street View

Samruan Wiangsamut, Chatklaw Jareanpon, and Sasitorn Kaewman

Polar Lab, Faculty of Informatics, Mahasarakham University, Thailand

Abstract. The Google street View (GSV) is used for many applications such as using for driver decision support system, collecting and merging the image for special purpose. The interactive system can be created by GSV. However, the most difficulty is the great image volumes. Moreover, for the reality, the map of interactive system ought to like with the physical place. This research proposed the management system created by the simple concept of database and designed by the simple color menu concept. Additionally, the map is constructed by the adjacent in the directed graph theory. This concept will demonstrate with the interactive system of the small pagoda in Thailand. The administrator management is easy and simple to manage this system, and the user or visitor feels like stay in the physical place and attracts to go to that place with more satisfied contentment.

Keywords: Web-based Management System, Interactive representation system.

1 Introduction

Google Street View (GSV) is developed by Google team from 2007. The Google's car is consisted of camera, laser and GPS [6]. The multiple camera will capture the multiple image at the same time. The image processing algorithm used to merge the multiple images to the 360-degree panoramic image. The GPS used to locate the picture. The interactive system of street view is made by constructing the map of multiple 360-degree panoramic with GPS. Many applications and researches interested in GSV such as Sun Surveyor shines with Street View [13], Pine processionary Moth assessing species distribution [11], and Air-ground matching for capturing the city [12]. Additionally, the GSV is able to create the interactive application that the user or visitor is able to walk though the virtual place. However, the image making the interactive system is plentiful. Moreover, the most difficulty is the relative direction and position between image.

This research proposed the new and simple management system to administrate the plentiful images from low cost single camera with special lens. Additionally, the graph theory is used for constructing the map in case of non GPS or tiny areas.

© Springer International Publishing Switzerland 2015
D. Camacho et al. (eds.), *New Trends in Computational Collective Intelligence,*
Studies in Computational Intelligence 572, DOI: 10.1007/978-3-319-10774-5_17

2   Literature Review

This research divided the literature review into 2 parts that are researches in Panoramic Picture with Camera, and researches in Google Street View and its application.

2.1   Researches in Panoramic Picture

A panoramic image is a wide view up to a full 360-degrees. A panoramic image can be captured from different ways such as:

1) By a single camera and a special lens or some kind of a mirror.

2) By multiple cameras, or one rotating camera with the merging image processing method.

This research interested in low cost of capturing process, then the single camera with special lens or rotating camera are reviewed. Shmuel and Moshe proposed the stereo panorama with a single camera [1]. This research generated a stereo panoramic image using circular projections from images by a single rotating camera. Peter and Franc used the concept of the rotating camera for panoramic depth imaging [2] with the rotational arm around the vertical axis. Apart from the rotating camera, David proposed the two binocular optical images using the data from single camera [3]. Moreover, this research proposed the Fourier transform and the Fourier shift to estimate the position of picture in stereo vision. It is very low cost, effort, and simple stereo-based applications. Jiang et al. proposed the panoramic 3D reconstruction using single camera together with the two planar mirrors [4]. Gurunandan and Shree proposed the Cata-fisheye camera for panoramic imageing using the fisheye lens capturing the reflection from the curved mirror [5]. This paper tested and evaluated in the optical quality of the captured image with high resolution panoramic images (3600*550 pixels) with a 360 degrees.

From these literature review, the single camera with the special lens is very low cost, and simple to collect the image with in single shot. The next literature review subsection is the literature reviews in Google Street View that is the well-known API to create the 360-degree panoramic-interactive system.

2.2   Researches in Google Street View

Google Street View (GSV) is to organize to world's information and user or visitor can access and use via internet [6] and [8]. This technology captured the image from the hundred of cities in different countries around the world. The multiple camera mounted on the car's roof such as R7 Street view consisted of 15 small rosetted cameras. Moreover, the Google Street View API allowed the developer to send the 360-degree panorama to generated the street view. From the large volume of images problems, many research interested in the GSV. Arturo and Serge proposed the method that removes the pedestrians from GSV image using SIFT, and RANSAC algorithm. The result is very clear in selected image, but still found problem in general outdoor scenes [15]. Victor and Chun-Ting proposed the development system for data collection in Location-based services by adding the automatic extraction and matching techniques for points of interest (POI) [7] and [9]. Jan and el al. proposed the detection of the sign recognition from GSV image for driver assistance systems [10]. Jerome and el al. proposed the Pine Processionary Moth (PPM) inferred from visual records derived from the panoramic views from GSV [11]. This research survey the Pine and insect species. Andras et al. proposed the Micro Aerial Vehicle Urban Localization from Google Street View Data [12]. This research combines the information from the GSV and air Vehicle called air-ground matching algorithm.

From all literature reviews, the GSV is used in many application such as decision support system, tree data collection, and also interactive application. However, the difficulty used the GSV is the constructed map from a lot of images and the management of collecting the images. This research is interested in the management of the image used for GSV and interactive application.

3   Our Proposed Method

Our proposed method is divided into 3 subsections that are capturing the image, Creating the map from the directed graph setting, and Designing the database.

3.1   Capturing the Image

Many ways are able to capturing the 360-degree panoramic image such as multiple camera, single camera with special device. For low cost, this research selects to use the single camera with the reflector. The 115 degree Vertical Field of View - 360 degree Horizontal Field of View will capture within one shot. This research uses the special lens from [14] as shown in the Figure 1(a). The reflection image is captured by the reflection mirror on the top of the lens. After that, the image processing will transform the reflected image to the 360-degree panoramic as shown in Figure 1b.

3.2   Creating the Map from the Directed Graph Setting

The interactive system is produced from the map of relation of the 360-degree panoramic images. The direction of the image is shown in Figure 2a. From the image A, the direction is anticlockwise, which the direction (heading angle) between A and B is set to 0, A and C is 90, A and D is 180, and A and E is 270. For the reality, the map must be constructed as same as the physical place. For example, from Figure 2b, this map is able to use the graph theory in term of adjacent node of the directed graph. The start picture is node 1, and the heading angle is shown with the dash line. The adjacent node of node 1 is node 2 with 0 heading angle. For easy and simple management, the adjacent node will store in the database with the heading angle. The next subsection will design and expend the database for constructing the map.

(a) Panoramic Optic Lens and Reflection 360-degree image

(b) 360-degree panoramic image

Fig. 1. Lens, Refection image, and 360-degree panoramic image: [www.0-360.com and www.panoguide.com]

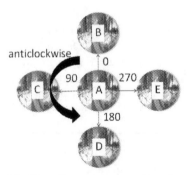

(a) The direction connection between image

(b) The map with direction

Fig. 2. The direction between the image and the directed map

3.3   Designing the Database

This research designs the database for manage the 360-degree panoramic picture and the relation between them via website or application. The database consisted of at least two tables that are 1) Image table and 2) Relative table as shown in Figure 3. The image table collects the normal information of the picture for example ID, url. The relative table collects the relation between the picture for example the picture 1 related with the picture 3 with 90 heading angle. Additionally from this management system, the picture 3 related with the picture 1 with 90 heading angle, will automatically insert to the database. It is easy and simple to manage the map of the interactive system via website as shown in Figure 4 and 5. Figure 4 shows the page that is able to add, upload, delete the 360-degree panoramic image. The designed page is very simple to understand with color menu: Add button with green color, Upload button with blue color, and Delete button with red color. Figure 5 shows the page that is able to add the connection between the 360-degree panoramic for constructing the map for example the Image ID336 connected with the Image ID 420 with 360 heading angle and ID 337 with 90 heading angle. From this design, the program is simple to send to the Google Street View API as shown in Figure 6. The simple code is very easy constructed by switch case or if. The picture 475 is connected with picture 474, 476 with heading angle 180, 360, respectively. The panoTile class is used for calling the arrow that makes the direction of connection in the map as shown in Figure 7.

Image table

Attribute	Data type	Data size	Detail	Example	Comment
id	int	-	Picture ID	1	PK
url	varchar	200	Picture Name	Place of Pagoda	
point	int	-	Start Picture	adminSuper	
p_code	varchar	300	ID of Place	35r45fb7b4n7h5n	
comment	varchar	250	Comment	Big Pagoda	

Relative table

Attribute	Data type	Data size	Detail	Example	Comment
id	int	-	Picture ID	1	FK
idreference	int	-	Connected Picture ID	3	
radius	int	-	Relative Degree Picture	90	

Fig. 3. Designing database

Fig. 4. Adding the picture via website

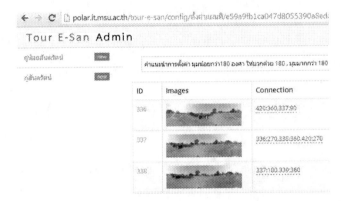

Fig. 5. Adding the connection via website

```
function getCustomPanorama(pano, zoom, tileX, tileY) {
        var panoTile = defaultPanoTile();
        panoTile.location.pano = pano;
        switch (pano) {

                case "474":

                        panoTile.links.push(linkWithId("521",90));

                        panoTile.links.push(linkWithId("475",360));

                        panoTile.links.push(linkWithId("553",270));

                        break;

                case "475":

                        panoTile.links.push(linkWithId("474",180));

                        panoTile.links.push(linkWithId("476",360));

                        break;
        }
        return panoTile;
}
```

Fig. 6. The simple code to create the interactive system using Google Street View API

Fig. 7. The interactive system using Google street view API

4   Experimental and Result

This research tested the proposed method in the pagoda in Thailand interactive system as shown in Figure 7. The tested interactive system shows at http://polar.it.msu.ac.th/pagodavr/. This interactive system is tested on various browsers and devices as shown in Table 1. This system interacts with the user by walking in the right way, non suspension, and fast display. Additionally, the human assessment report is tested on the interactive performances as shown in Table 2.

Table 1. Browsers and devices tested performance

Browsers and devices	Result
iOS operating system from iPad and iPhone	
Safari	Pass
Chrome	Pass
Dolphines	Pass
Mercury	Pass
Android operating system	
Chrome	Pass
Dolphines	Pass
Opala	Pass

In the viewpoint of the image management system, the human assessment report tested as shown in Table 3.

Table 2. Interactive performance contentment

Interactive performance topics	Totally disagree	Disagree	Average	Agree	Totally agree	Mean
The web page is beautiful, appropriated and attractive .	1	4	5	11	7	3.71
Feel like in the physical place	0	1	8	16	3	3.78
Fast display and no twitch	0	3	7	16	2	3.64
Resolution of the display picture	2	1	6	14	5	3.71
Attractive going the physical place	0	1	5	15	7	4.03
Content and fun more than normal web	1	4	1	17	5	3.78
New and Modern	1	2	4	16	8	3.92
Total	4	12	31	94	30	3.81
Interpretation of results	**This web site is more satisfied.**					

Table 3. Administrator of management contentment

Interactive performance topics	Totally disagree	Disagree	Average	Agree	Totally agree	Mean
The image management system is easy to add image .	0	0	4	7	9	4.25
The image management system is easy to make the direction of image	1	1	3	10	5	3.85
Fast operations such as adding,editing and deleting	0	0	2	3	15	4.65
Easy to install the image management system	0	0	1	6	13	4.60
Easy to understand the instruction of the image management system	0	0	4	6	10	4.30
The color design is helpful	0	0	0	9	11	4.55
Total	1	1	14	41	63	4.36
Interpretation of results	**This web site is most satisfied.**					

5 Conclusion and Future Work

This research designs and develops the management system for the interactive system using Google Street View API. This management is very simple to add, delete, update the image via website with the simple color menu. Additionally, this management website is able to construct the map and the connection between the image. That API is able to display in various devices. From human assessment report, the users feels like in the physical place and attracts going to that place.

Additionally, the administrator image management system is fast operations such as adding, editing and deleting and easy to understand the instruction.

However, the remain problem of the reality is the direction of image when the user back to the previous image, the user must see the difference view. The heading angle will calculated instead of using the constant value.

Acknowledgement. This research received funding and is supported by The National Research Council of Thailand (NRCT) and Mahasarakham University.

References

1. Peleg, S., Ben-Ezra, M.: Stereo Panorama with a Single Camera. In: IEEE Computer Society Conference on Computer Vision and Pattern Recognition, pp. 395– 401 (1999)
2. Peer, P., Solina, F.: Panoramic Depth Imaging with a SIngle Standard Camera. International Journal of Computer Vision 47, 149–160 (2002)
3. Rodriguez, B., et al.: Interactive Design of personalised tourism routes. Tourism Management 33, 926–940 (2012)
4. Wei, J., et al.: Panoramic 3D Reconstruction Using Rotating Camera with Planar Mirrors. In: The 6th Workshop on Omnidirectional Vision, Camera Networks and Non-classical cameras, pp. 1–8 (2005)
5. Krishman, G., Nayar, S.K.: Cata-Fisheye Camera for Panoramic Imaging. In: IEEE Workshop on Applications of Computer Vision, pp. 1–8 (2008
6. Dragomir, et al.: Google Street View: Capturing the World at Street Level. IEEE Computer Society, 32–38 (2010)
7. Tsai, V.J.D., Change, C.-T.: Feature Positioning on Google Street View Panoramas. ISPRS Auuals of the Photogrammetry, Remote Sensing and Spatial Information Sciences 1-4, 305–309 (2012)
8. Google Street View Imape API,
   https://developers.google.com/maps/documentation/streetview/
9. Tsai, V.J.D., Change, C.-T.: Three-dimensional Positioning from Google Street View Panoramas. IET Image Processing, 1–11 (2012)
10. Salman, J., et al.: Google Street View Images Support the Development of Vision-Based Driver Assistance Systems. In: IEEE Intelligent Vehicles Symposium, pp. 891–895 (2012)
11. Rousselet, J., et al.: Assessing Species Distribution Using Google Street View: A Pilot Study with the Pine Processionary Moth. PLoS ONE 8(10), 1–7 (2013)
12. Majdik, A.L., et al.: MAV Urban Localization from Google Street View Data. In: IEEE/RSJ International Conference on Intelligent Robots and Systems (2013)
13. Ratana, A.: Sun Surveyor shines with Street View,
    http://googlegeodevelopers.blogspot.com/2013/09/
    sun-surveyor-shines-with-street-view.html
14. Panoramic Optic, http://www.0-360.com/
15. Flores, A., Belongie, S.: Removing pedestrians from Google Street View Images. In: IEEE Computer Society Conference on Computer Vision and Pattern Revognition Workshop (CVPRW), pp. 53–58 (2010)

Part IV
Intelligent Computational Methods

Simple Gamer Interaction Analysis through Tower Defence Games

Fernando Palero, Antonio Gonzalez-Pardo, and David Camacho

Computer Science Department,
Universidad Autónoma de Madrid, Spain
fernando.palero@inv.uam.es, {antonio.gonzalez,david.camacho}@uam.es
http://aida.ii.uam.es

Abstract. In the last years, the Video Game industry has growth considerably, capturing the attention of the research community. One of the research hot topics in videogames is related to the identification of gamers behaviour while they are playing the game. This work presents an initial case related to the identification of users behaviour in a particular kind of videogame through gamer interaction extraction and analysis. The Video Game selected in this work is a *Tower Defence Game*, called *OTD*, where the user needs to build towers, in a 2-D grid, to avoid the enemies to reach the exit point of the level. It has been created a framework that allows extract the information from the game and later use statistical techniques to analyse the gamers behaviour. Finally, some experiments have been carried out to test this framework.

Keywords: Player Behaviour, Video Games, K-Means, Tower Defence Game.

1 Introduction

Nowadays a wide number of Computer Science researchers are focused on intelligence Video Games [1,13,17,19,22], design and development. Several techniques and methods from areas such as Artificial Intelligence (AI) or Data Mining (DM) have been applied to gamers behaviours analysis [3,16], to generate intelligent enemies [22,24], or to imitate the human behaviour [11,14], among others. Maybe, one of the most known applications is related to the development of controllers to automatically define real behaviour of Non-Player Characters (NPC). In this topic, there are several works focused on really famous games such as *Ms. PacMan* [7], *Physical Travelling Salesman Problem* (PTSP) [15], *Super Mario Bros.* [12] or *Starcraft* [20]. Other works have been focused on the generation of automatic levels [18] or the validation of those levels by finding the different paths that reaches the exit [8,9].

This paper presents an initial case study related to the identification and analysis of gamers behaviours in Video Games using visualization techniques to understand what happened during the game execution. For example, the ESOM [4] visualization technique has been used to identify the clusters of player patterns in the famous game *Tomb Raider Underworld* [4]. Other example can be found in [6] where authors study the user behaviour in the game *Kane & Lynch and Fragile Alliance* by the use of diagrams to

© Springer International Publishing Switzerland 2015
D. Camacho et al. (eds.), *New Trends in Computational Collective Intelligence*,
Studies in Computational Intelligence 572, DOI: 10.1007/978-3-319-10774-5_18

understand the reasons of why the players die and map levels visualizations to represent the paths followed by the players.

The platform designed to achieve our goals is based on an open-source *Tower Defence* game[1], called *OTD*, that it is a subgenre of real-time strategy Video Game. This analysis is based on previous works [2,10] where the interaction of users were automatically extracted and analysed from an Educational Virtual World to differentiate the different student communities based on the interactions among them.

The goal in this kind of game is to avoid that a set of enemies, that appear in different waves, reach the exit of the level. In order to do that, users need to build in the map different traps, or some defensive buildings, that difficult enemies to cross the map. In this initial work, the players behaviour is analysed by taking into account only the position of the different towers placed by the user in a 33×60 grid. Two different kind of experiments have been carried out in this work. In the first one, the different waves of enemies, that try to reach the exit of the level, has a constant size (the number of enemies in each wave does not change). In the second experiment, the number of enemies in each wave will linearly grows making the users to take more decisions as the game progresses.

The remainder of this paper is organized as follows: Section 2 provides some background and related work on game analysis. Section 3 presents a platform for game data extraction and analysis for the Tower Defence Game considered. Section 4 presents some experimental results.Finally, Section 5 shows the conclusions and future lines of work.

2   Related Work

The number of players playing on-line games generate a huge quantity of data that can be later analysed. These data can be used to remote monitoring users behaviour, to analyse game servers, mobile devices, or to predict users preferences. There is a large number of works related to these issues. Next, a brief description of some related works are seen.

D. Anders et al, [5] analysed the player behaviours in *Tera* and *Battlefield 2* games. Tera is a Massively Multiplayer Online Role-playing Game (MMORG) with features such as questing system, crafting, player vs. player action, as well as an integrated economy. The dataset of this game contains 250.00 player characters with two groups of features. One of them is relate to the character ability (class, race, strength, attack, etc.), and the other is composed by the gameplay features (monsters killed, in-games friends). Battlefield 2 is a First-Person Shooter (FPS) game with strategic and tactical elements reflecting small-scale warfare. The datasets available at P-stats server[2], and contain about 10.000 features from 69.313 players characteristics. In these games players have not any special abilities, so the data recorded represent gameplay features: Score, Skill, Kill/Death Ratio, Round played, etc. Tera and Battlefield 2 datasets had been analysed using clustering algorithms, K-Means and SVM, to detect players behaviours.

---

[1] Available at: http://sourceforge.net/projects/otd/

[2] http://p-stats.com/

In Thompson [21], the authors use the interaction data collected from 3.360 Star Craft 2 players from 7 level of expertise, ranging from novices to full-time professionals. The aim of this study is to create a model to predict the behaviour of expertise players. Because of the complexity of Real Time Games for gathering data, it is necessary to use predictor variables that relate to cognitive-motor abilities, cognitive load variables and variables that measure the amount of resources required to perform the task. The predictor variables, cognitive load variables and variables that measure the amount of resources have been used in the algorithm random forest to classify the player behaviours.

In Tomb Raider[4], the goal of the research is to create a better AI which capture the interest of the players. The data has been taken from 1365 users, which completed all the game levels. From the dataset six statistical features were extracted that correspond to high-level playing behaviours: completion time, number of death, causes of death and help on demand. These features have been analysed using k-means and ward hierarchical methods.

The research show in Yannakakis[23] introduces an effective mechanism for obtaining computer games of high interest (i.e. players satisfaction). The proposed approach is based on the interaction of a Player Modelling (PM) tool and a successful on-line learning mechanism based on prey/predator computer games.In this study, Bayesian Networks(BN) were applied, trained on computer-guided player data, as a tool for inferring appropriate parameter values for the chosen on-line learning (OLL) mechanism. The results obtained shown that PM positively affects to the OLL mechanism generating games of higher interest for the player. In addition, this PM-OLL combination, in comparison to OLL alone, demonstrates a faster adaptation in challenging scenarios of frequent changing playing strategies.

3   Tower Defence Framework Architecture

This section presents a framework architecture based on a open-source Tower Defence platform (see Figure 1) that have been developed to study the gamers behaviours. The framework has been designed using three different modules. The Wave Generator Module (WGM) is the responsible to generate a fix number of variable hordes of enemies in each wave. The Data Recovery Module (DRM) allows to automatically extract data from the game platform, and gather the interaction from the users. Finally, the Computational Intelligence Module (CIM) analyses and returns the distribution of the gameplays. These distributions have been analysed with visualization techniques (histograms) to determine the strategies.

The WGM determines the size of the horde in a wave. The equation 1 is used to define the enemies number of the horde that will be generated in a gameplay. Where N represents the number of enemies, W represents the current wave and α is a growing factor for the enemies generation. For $\alpha = 0$, the number of enemies in the horde is constant in each wave. If $\alpha = 1$, the amount of enemies grows linearly and, finally, if $\alpha = 2$ the horde growth has quadratic grows factor.

$$\#Enemies = NW^{\alpha} \tag{1}$$

The *DRM* extracts data from the Tower Defence game which is pulled in two different categories, Game Data (GD) provides information from the environment, the environment information is based on those features predefined by the game (i.e. the number of waves in a game, the type of the enemies, the size of an horde, etc). And Interplay Data (ID): there are two types of Interplay Data, Player Interaction (PI), and Game Snapshot (GS). PI data represents the actions of the player, and the events of the game during the execution and GS providing a detailed description about state of the game for each instant of time. Both kind of IP data (PI and GS) are gathered from the platform every second during a gameplay.

Finally, the *CIM* is the responsible to carry out the analysis of the data gathered by *DRM*. It works based on two basic processes. The first process recovers the tower positions from the databased, and the second process is used to calculate both, the distributions of X, Y, and the distribution of euclidean distances from the towers to the entry point. Using these features an initial classification of gamers strategies are generated.

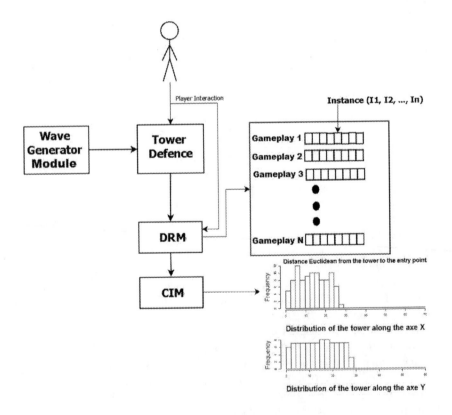

Fig. 1. Framework Architecture based on the OTD game platform

4   Experimental Results

In this section we discuss the different types of strategies that have been used by the players, the strategies have been used to classify the different behaviours. In this initial approach, we use where the towers are placed on the game board according to the type of the waves generated to identify the player strategies. The board is two-dimensional grid where the X dimension represents the width and the Y represents the height (the size of the X is 60 and the size of Y is 33), the starting point of the enemies is located in the position $(0, 17)$ and the exit point is located in $(60, 17)$. The size of the grid is important, when more larger the grid more space must be protected and then we can apply a huge number of strategies. For this reason we choose a huge grid of 33×60.

In the different waves of a game, only one horde of enemies is generated. The enemies of these hordes have the same stamina and strength, but the number of enemies could change during the gameplay. Changing the size of the hordes, two different experiments have been carried out (constant and linear) to determine the user strategies.

Any wave generated has been divided into two different phases. The first phase has no enemies and the gamer has 5 seconds to put or remove towers in the board. During the second phase the enemies will try to reach the exit, so the user will need to put more towers, remove the existing ones or placed new towers in other parts of the board. In this second phase the enemies will appear with a frequency of 0.5 seconds, the number of enemies will be generated according to equation 1. Once both phases have conclude there are only two possibilities, a successful strategy (if the player is able to destroy all the enemies), or a failed strategy (if at least three o more enemies are able to reach the exit).

The dataset has 24 games, 12 of them with horde size constant in each wave and the other 12 with linear growth of the horde size. In the experiment the gamers, which play for first time the Tower Defence Game, have been divided in two groups. In the first group, the players have been assigned with a particular strategy (therefore this player must follow a predefined strategy to place towers in the grid). And in the second group, gamers play without any assigned strategy. Each gamer plays three times in the first experiment, and three times in the second. Once the first group of strategies have been analysed, the strategies are used to classify the strategies followed by the second group of players.

4.1   Strategies

Using the data extracted from the different gameplays we find four strategies in the first group:

- The first one, *Zigzag* distribution, this strategy is useful to slow the enemies. The goal is that enemies take them longer time to reach the exit point, and so the towers have more time to shot them, see figure 2. This strategy has the characteristic that is developed along the X dimension, in the Y dimension the towers are placed in a fixed rank of positions. This way of placing towers makes the three distributions are uniform.

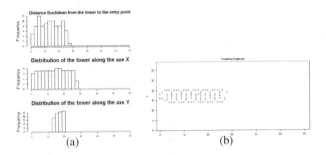

Fig. 2. Zigzag distribution

– The second one, *Vertical* distribution has the characteristic that towers are distributed along the dimension Y in one or two columns. This tactic concentrate the strength of the towers in a big column cluster to destroy enemies, see figure 3. In this strategy we can observe that towers are placed along the axis Y, this dimension has Gaussian distribution. Furthermore, the distribution on the X dimension has grouped the majority of the information in one bin, this happens because the coordinates X of the towers are all the same.

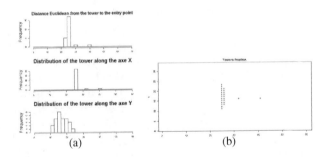

Fig. 3. Vertical distribution

– The third one, *Grouped* distribution joins towers in small clusters to increase the damage that enemies receive, these clusters are spread throughout the game board, see figure 4. The clusters can have different aspect: square, circular, lengthened, etc. This strategy has the characteristic that histograms have a saw distribution.
– Finally, *Horizontal* distribution creates a horizontal wall of towers across the game board. This strategy is similar to *Vertical* strategy, the aim is to concentrate the strength of the towers in a big horizontal cluster to destroy the enemies, see figure 5. This strategy has the characteristic that is developed along the X dimension, we can observe that the distribution of the dimension X and the distribution of the euclidean distance is uniform and the information from the dimension Y can be grouped in one bin.

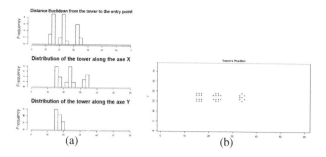

Fig. 4. Grouped distribution

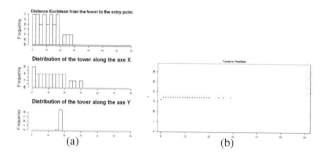

Fig. 5. Horizontal distribution

Once the players strategies have been correlated with their distributions, the distribution of the second group of games have been studied. Only 4 new players have been used to carry out this initial analysis. We have done a visual analysis, we have compared the distributions of the second group with the first group. From these new experiment, some conclusion can be given.

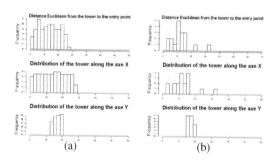

Fig. 6. Fig (*a*) shows the Zigzag distribution, whereas Fig (*b*) shows the distribution from one of the second group gamers

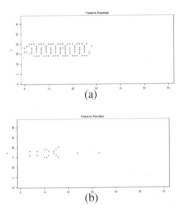

Fig. 7. Fig (*a*) shows how the player places towers using Zigzag distribution, whereas Fig (*b*) shows the towers position from the second group of gamers.

Horizontal and *Vertical* distributions are easy to identify. Where a *Horizontal* distribution is used, towers are concentrated in axis Y in one bin. However, in *Vertical* distributions, this effect appears in axis X. The same occurs with the *Grouped* distribution, which has a saw distribution in the axis X and Y that is easy to recognize. Finally, *Zigzag* distribution is difficult to identify because the three distributions are uniform and there are several ways to place towers that can create this kind of distribution. For example, in the *Horizontal* distribution, both the euclidean distance and the axis X, has an uniform distribution. It is easy to confuse both strategies. On the other hand, the figure 7 presents a game where the gamer does not use a *Zigzag* tactic but the distribution (see figure 6) is similar to *Zigzag*.

5   Conclusions and Future Work

This work provides an initial study on gamers interaction and their strategies used in a Tower Defence game. To achieve this it has been designed a framework based on an open-source Tower Defence platform. Two experiments have been carried out to study the players interaction: in the first experiment, the size of the hordes in each wave is constant and the second experiment the size of the hordes in each wave have a linear growth. From the data extracted from the game, we have gathered the tower positions that are placed during the game to calculate three features: the distribution of the towers in the X dimension, in the Y dimension and the distribution of the euclidean distances from tower positions to starting point. These features have been used to analyse the player strategies.

Using some initial experiments four different strategies have been detected. *Zigzag* strategy, *Vertical* strategy, *Grouped* strategy and *Horizontal* Strategy. *Zigzag* strategy has the characteristic that their three distributions are approximately flat. In *Vertical* strategy the towers are placed along the dimension Y, the dimension Y has a Gaussian distribution, and the other two distributions show the towers are mainly concentrated in one bin, because the towers are positioned in the same X position. *Grouped* strategy tend to follow a saw distribution. Finally, in the *Horizontal* strategy shows a distribution of the euclidean distance has a saw shape, the dimension X has a uniform distribution dimension Y concentrate most information in one bin.

As it was described in the section 4.1, the towers positions distribution is not enough to identify the players strategies. In future works, it will be necessary to use more features to perform a better classification and study techniques that could help to select the best game features.

Acknowledgement. This work was supported by the Spanish Ministry of Science and Education under Project Code TIN2010-19872 and Savier Project (Airbus Defence & Space, FUAM-076914). The authors are grateful for all the support obtained through Airbus Defence & Space, specially from its project members José Insenser, Juan Antonio Henriquez and Gemma Blasco.

References

1. Alayed, H., Frangoudes, F., Neuman, C.: Behavioral-based cheating detection in online first person shooters using machine learning techniques. In: 2013 IEEE Conference on Computational Intelligence in Games (CIG), pp. 1–8. IEEE (2013)
2. Berns, A., Gonzalez-Pardo, A., Camacho, D.: Game-like language learning in 3-d virtual environments. Computers and Education 60(1), 210–220 (2013)
3. Dey, R., Child, C.: Ql-bt: Enhancing behaviour tree design and implementation with q-learning. In: 2013 IEEE Conference on Computational Intelligence in Games (CIG), pp. 1–8. IEEE (2013)
4. Drachen, A., Canossa, A., Yannakakis, G.N.: Player modeling using selforganization in tomb raider: Underworld. In: Proceedings of the 5th International Conference on Computational Intelligence and Games, CIG 2009, pp. 1–8. IEEE Press, Piscataway (2009)
5. Drachen, A., Rafet, S., Bauckhage, C., Thurau, C.: Guns, swords and data: Clustering of player behavior in computer games in the wild. In: Proceedings of CIG 2012, pp. 163–170. IEEE (2012)
6. Drachen, A., Canossa, A.: Towards gameplay analysis via gameplay metrics. In: Proceedings of the 13th International MindTrek Conference: Everyday Life in the Ubiquitous Era, pp. 202–209. ACM (2009)
7. Gagne, D.J., Congdon, C.B.: Fright: A flexible rule-based intelligent ghost team for ms. pac-man. In: 2012 IEEE Conference on Computational Intelligence and Games (CIG), pp. 273–280. IEEE (2012)
8. Gonzalez-Pardo, A., Palero, F., Camacho, D.: An empirical study on collective intelligence algorithms for vide games problem-solving. Computing and Informatics (in press, 2014)
9. Gonzalez-Pardo, A., Palero, F., Camacho, D.: Micro and macro lemmings simulations based on ants colonies. In: Evostar. EvoGames (page in press, 2014)
10. Gonzalez-Pardo, A., Rosa, A., Camacho, D.: Behaviour-based identification of student communities in virtual worlds. Computer Science and Information Systems 11(1), 195–213 (2014)
11. Johansson, A., Dell'Acqua, P.: Emotional behavior trees. In: 2012 IEEE Conference on Computational Intelligence and Games (CIG), pp. 355–362. IEEE (2012)
12. Karakovskiy, S., Togelius, J.: The mario ai benchmark and competitions. IEEE Transactions on Computational Intelligence and AI in Games 4(1), 55–67 (2012)

13. Nguyen, K.Q., Wang, Z., Thawonmas, R.: Potential flows for controlling scout units in starcraft. In: 2013 IEEE Conference on Computational Intelligence in Games (CIG), pp. 1–7. IEEE (2013)
14. Polceanu, M.: Mirrorbot: Using human-inspired mirroring behavior to pass a turing test. In: 2013 IEEE Conference on Computational Intelligence in Games (CIG), pp. 1–8. IEEE (2013)
15. Powley, E.J., Whitehouse, D., Cowling, P.I.: Monte carlo tree search with macro-actions and heuristic route planning for the physical travelling salesman problem. In: 2012 IEEE Conference on Computational Intelligence and Games (CIG), pp. 234–241. IEEE (2012)
16. Rosenthal, C., Congdon, C.B.: Personality profiles for generating believable bot behaviors. In: 2012 IEEE Conference on Computational Intelligence and Games (CIG), pp. 124–131. IEEE (2012)
17. Schaul, T.: A video game description language for model-based or interactive learning. In: 2013 IEEE Conference on Computational Intelligence in Games (CIG), pp. 1–8. IEEE (2013)
18. Shaker, N., Togelius, J., Yannakakis, G.N., Weber, B., Shimizu, T., Hashiyama, T., Sorenson, N., Pasquier, P., Mawhorter, P., Takahashi, G., et al.: The 2010 mario ai championship: Level generation track. IEEE Transactions on Computational Intelligence and AI in Games 3(4), 332–347 (2011)
19. Sifa, R., Bauckhage, C.: Archetypical motion: Supervised game behavior learning with archetypal analysis. In: 2013 IEEE Conference on Computational Intelligence in Games (CIG), pp. 1–8. IEEE (2013)
20. Synnaeve, G., Bessiere, P.: A bayesian model for rts units control applied to starcraft. In: 2011 IEEE Conference on Computational Intelligence and Games (CIG), pp. 190–196. IEEE (2011)
21. Thompson, J.J., Blair, M.R., Chen, L., Henrey, A.: Video game telemetry as a critical tool in the study of complex skill learning. PLoS One 8(18), 1–12 (2013)
22. Traish, J.M., Tulip, J.R.: Towards adaptive online rts ai with neat. In: 2012 IEEE Conference on Computational Intelligence and Games (CIG), pp. 430–437. IEEE (2012)
23. Yannakakis, G.N., Maragoudakis, M.: Player modeling impact on player's entertainment in computer games. In: Ardissono, L., Brna, P., Mitrović, A. (eds.) UM 2005. LNCS (LNAI), vol. 3538, pp. 74–78. Springer, Heidelberg (2005)
24. Young, J., Smith, F., Atkinson, C., Poyner, K., Chothia, T.: Scail: An integrated starcraft ai system. In: 2012 IEEE Conference on Computational Intelligence and Games (CIG), pp. 438–445. IEEE (2012)

Markov Logic Network Based Social Relation Inference for Personalized Social Search

Junho Choi[1], Chang Choi[2], Eunji Lee[2], and Pankoo Kim[2,*]

[1] Division of Undeclared Majors, Chosun University,
309, Pilmun-daero, Donggu, Gwangju, South Korea
xdman@chosun.ac.kr
[2] Departmrent of Computer Engineering, Chosun University,
309, Pilmun-daero, Donggu, Gwangju, South Korea
{eunbesu,enduranceaura}@gmail.com, pkkim@chosun.ac.kr

Abstract. Most recommendation systems are based on historical data and profile files. The most common method is collaboration filtering. After analysis, a collaboration filtering recommender system differentiates one user profile from another to determine what to recommend. There has been substantial research on personalized social search. However, previous research has neglected semantic social information, making no use of definite relations between objects. This problem can be solved using ontology and inference rules. In this paper, Markov-logic-network (MLN)-based social relation inference is performed using social user information, such as country, age, and preference. In addition, this paper evaluates whether the inference results regarding social relations have been correctly predicted based on social user data. The user's personal and business relations are inferred based on MLNs and a social network comprised of user profile data.

Keywords: Markov Logic Networks, Social Network Services, Relation Inference.

1   Introduction

Online social media have become popular tools for content sharing and relationship maintenance. People create digital identities to distribute videos or photos, share opinions about books and movies, or just maintain friends' contact information. There are many incentives for users who contribute to these social networks, but the addition of social information to online databases has dramatically increased their popularity. The popularity of these collaborative systems has resulted in a substantial increase in user-generated content and metadata [1].

Most recommendation systems are based on historical data and profile files. A system analyzes where users have visited previously, what they viewed, and how they rated it. The most common method is collaboration filtering. After analysis, a collabo-

---

* Corresponding author.

© Springer International Publishing Switzerland 2015
D. Camacho et al. (eds.), *New Trends in Computational Collective Intelligence*,
Studies in Computational Intelligence 572, DOI: 10.1007/978-3-319-10774-5_19

ration filtering recommender system differentiates one user profile from another to determine what to recommend. This approach suffers from the data sparseness and first rate problems. Other recommendation systems are content based. A content-based system provides recommendations for users by comparing the content representations contained in a list. This kind of recommendation requires more effort to extract list content. Users of such a system do not need to provide significant feedback. On the other hand, it has the drawback of less interaction between the system and the user [2, 3].

Social network analysis views social relationships in terms of network theory; a network consists of nodes (representing individual actors within the network) and edges (which represent relationships between the individuals, such as friendship, kinship, organizational, or sexual relationships) [4].

There has been substantial research on personalized social search. However, previous research has neglected semantic social information, making no use of definite relations between objects. This problem can be solved using ontology and inference rules [5, 6, 7].

In this paper, Markov-logic-network (MLN)-based social relation inference is performed using social user information, such as country, age, and preference. In addition, this paper evaluates whether the inference results of social relations have been predicted correctly based on social user data. The user's relations consist of inferred personal and business relations based on Markov logic networks and a social network comprised of user profile data.

2   Related Work

2.1   Semantic Social Information Retrieval

Recently, the fields of information retrieval and social network analysis have been bridged by social information retrieval (SIR) models, which extend conventional information retrieval (IR) to incorporate the social context of search and recommendation. One aspect of SIR is the modeling of the trust and reputations of the members of a social network, which is then typically used for enhancing recommendation systems [8]. The SIR method considers common interest and user social network structure. SIR is a type of web search that takes into account varying sources of metadata, such as collaborative discovery of web pages, tags, social ranking, commenting on bookmarks, news, images, videos, knowledge sharing, podcasts and other web pages [9]. In addition to information retrieval, data analysis in social environments has aroused considerable interest from closely related research fields, such as data mining, sentiment analysis, and machine learning. Novel research directions (e.g., social computing, user behavior modeling, and collective intelligence) have increasingly emerged to play important roles in modern information analysis [10].

2.2   Markov Logic Network

This section is an excerpt from our previous paper [11]. Probability is the yardstick for expressing degrees of certainty regarding statistical phenomena, while probability theory is a mathematical description of non-probabilistic phenomenon that can be expressed in a graph model.

Probabilistic reasoning is an area that aims to find efficient mechanisms for reasoning under uncertain knowledge expressed through probability theory. Probabilistic graphical models provide compact and expressive tools for dealing with uncertainty and complexity by joining concepts from probability theory and graph theory in the same representation. Bayesian networks and Markov random fields can be used to determine the probability of an inference result [11, 12].

A Markov network, also called a Markov random field or undirected graphical model, is a discrete time probability process that has Markov properties. Markov networks were introduced to compensate for Bayesian networks' lack of circular relations [11].

A Markov network expresses dependent relations similarly to a Bayesian network. Markov networks represent a joint probability distribution of random variables as an undirected graph, where the nodes represent the variables, and the edges correspond to direct probabilistic interactions between neighboring variables. This interaction is parameterized by potential functions. There is a potential function for each clique (i.e., completely connected sub-graph) in the graph, with a potential function being a non-negative real-valued function of the state of the corresponding clique [11].

A Markov network can be expressed using a log-linear model. A potential function can be replaced by the function that represents the sum of exponentiated weights, as shown in Equation 1 below.

$$P(X = x) = \frac{1}{Z} \exp \left(\sum_i w_i f_i(x) \right)$$

In fact, the partition function Z is an important function that includes many statistical concepts. The function is expressed using probability values between zero and one, as it is transformed in various ways to resolve problems. This means that constraint conditions are in effect.

3   MLN-Based Social Relation Inference

3.1   Collection of Social User Information

Social user information is classified as user profile data and tag data for use in personal recommendation information. User profile data consist of keywords for social content expression, such as country, age, and preference. These kinds of feature are used to create a friend-of-a-friend (FOAF) profile through an openAPI in social network services (SNS). Table 1 provides a representation of ontology properties for user profiles. The user profile data are used for relations and recommendations among users.

Table 1. Ontology property for user profile data

User Profile Data	Properties
Name	foaf:firstName
Last name	foaf:lastName
Username	foaf:id
E-mail	foaf:email
Gender	foaf:gender
Birthday	bio:birthday
Interest	foaf:interest
Role	org:role

Tag data are easily moved to the desired content because there can be direct connections between tag data and social content. In addition, there can be confirmation of another user's contents, as well as sharing of links between users using tag data. Recently, social tags have been used to gather and manage various kinds of SNS information, because social content topics can be confirmed easily using tag data. Finally, tag data have become the basic information for sharing common interests. In this paper, tag data consist of basic and associative relation information. Table 2 provides a representation of ontology properties for tag data.

Table 2. Ontology property for tag data

Tag Data	Properties
Contents	tag:contents
Subject	tag:subject
Document Type	tag:documentType
Tasks	tag:tasks
Opinion	tag:opinion
Purpose	tag:purpose

3.2   Extraction of Social Relation

This paper aims to use MLNs, which combine Markov networks with logic, with a view to expressing inference results regarding ontology probabilistically based on social relation extraction. In other words, a new social relation is extended using MLNs. If a relation sequence for social relation extraction has been referenced in multiple social domains, the shortest path is used. A relation sequence is a set of adjacency relationships between two nodes, such as x and y, in network. The ranking of extracted relations is determined to check connections between instances, where an instance is the most important concept of social user information. In addition, the ranking of relations is determined based on social user information such as popularity, relation frequency, and preference.

The relation distance is the number of links between nodes of social users in SNS. The popularity of a social user is determined based on social user information between a user and another user, such as the number of relations and the intensity. A user who maintains varied relationships with other users has substantial social impact. Therefore, the social popularity of nodes in a relation sequence is used as a social effect for the ranking of extracted relations.

The relation frequency is the number of relations. A path of high relation frequency receives a high ranking, as a result of receiving substantial information. Preference is the main criterion for the weight value of a user path.

Table 3. Relation property for social user

FOAF	Relationship	Organization
Knows Member	Employed By Employer Of Friend Of Has Met Influenced By Mentor Of Works With	Member

3.3   Social Relation Inference Based on MLNs

This paper aims to extend new relations for social relation inference using an MLN algorithm based on SNS connection relations. An MLN L is a pair (F_i, w_i), where F_i is a formula of first order logic, and w_i is a weight value. Weight values are provided in MLNs to clarify differences in result values for classification. There is no special standard for weight. Generally, when weight value increases, probability value increases, along with an increase in restrictions. Since logarithms are used, the deviation depending on weight value is small [4]. $M_{L,C}$ is defined using a Markov network and a finite constant set $C = \{C_1, C_2, \ldots, C_{|c|}\}$. $M_{L,C}$ has the feature that all ground instances and feature values are true (1) or false (0). w_i is the weight value of a feature. A Markov logic inference is calculated as the sum of weight values using true formulas based on evidence.

The MC-SAT algorithm, which is used in MLNs, performs learning based on Markov Chain Monte Carlo algorithms. The Markov Chain Monte Carlo algorithms enable extraction of a sample at a specific stage before clarifying differences in weight values within a margin of error within which extracted samples can be accepted [4]. In this paper, learning is performed based on connection relations using the MC-SAT algorithm.

The learning of a Markov network for new relations is processed by each group after extraction of a user's profile information and social relations. For example, if there is a social relation between users x and y, then these two users have an intimate relationship.

Table 4. Inference Rules based on Social Relation

Weight	Formula
0.07	$\forall$ x,y,z : relation(x, y) $\lor$ relation(y, z) $\rightarrow$ friend(x, z)
1.89	$\forall$ x,y : friend(x, y) $\leftrightarrow$ friend(y, x)
1.02	$\forall$ x,y,z : hasInterest(x, z) $\lor$ hasInterest(y, z) $\rightarrow$ friend(x, y)

The probability that a predicate is true is inferred in an MLN after extension of social relations between previous and new social relations. The user's semantic relation inference consists of SNS instances and domain rules. This paper proposes the most popular connection path between two users through MLN-based probabilistic inference. The highest popularity connection path between two users is a new connection relation using probabilistic inference.

Table 5. Classification Rules based on Social Relation

Weight	Formula
0.09	$\forall$ x,y,z : friend(x, y) $\lor$ worksWith(x, z) $\rightarrow$ worksWith(y, z)

4   Experiment and Evaluation

In general, inference for ontology is performed using an ontology inference engine. There are a variety of inference engines such as Jena, FOWL, Pellet, and Fact++. These inference engines determine simply if there is a result from an inference or not. Recently, probabilistic inference methods such as MLNs have emerged and classified problems that cannot be defined easily in a probabilistic way, providing better results. Against this background, this paper aims to use a probabilistic inference method based on MLNs, rather than existing ontology inference engines, for conducting experiments [4].

This paper evaluates whether the inference results of social relations have been predicted correctly based on 12 social user data. The user relation is inferred personal and business relations based on MLNs, and the social network is comprised of user profile data. Figure 1 is a social network based on personal and business relations.

The accuracy of social relations in MLNs is evaluated using a comparison and analysis based on user business relations. Business relations are represented as directional dotted lines of social relations between objects, such as the foaf:knows property. If there is a connection relation representing a business relation, then the distance is one.

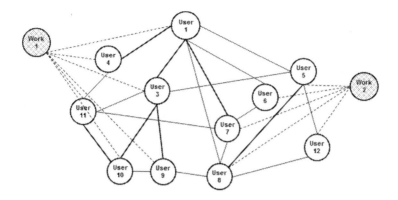

Fig. 1. Social Network based on User Profile Data

In addition, a social network has the foaf:knows property, if it is represented by the same business relation. Finally, a new inferred semantic social relationship is represented by an appropriate social relation. In this paper, learning is performed based on connection relations using the MC-SAT algorithm, as shown in Tables 4 and 5.

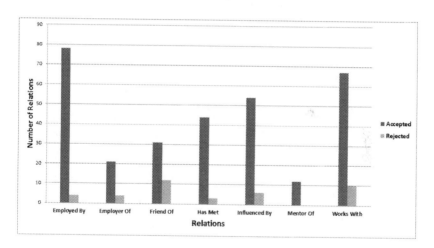

Fig. 2. Comparison of "Accepted" with "Rejected" by Relation Inference

Figure 2 is a comparison of created social relations by relation type in MLNs. The "Accepted" result is 85%, and the "Rejected" result is 15%. Table 6 shows the accuracy rate by user social relation, using the results of Figure 2.

Table 6. Accuracy Rate by User Social Relation

Inference Type	Accuracy Rate
Personal Relation	75.7%
Business Relation	61.2%

5   Conclusion

Previous research conducted on social networks has often emphasized social network architecture, that is, the existence of paths between objects in a network and the centrality of the objects in it. However, studies on semantic association in the network are rare. Studies on searching semantic associations between entities are necessary for personalized social search. In this paper, MLN-based social relation inference was performed using social user information, such as country, age, and preference. In addition, this paper evaluated whether the inference results of social relations have been predicted correctly based on 12 social user data. The user relation is inferred personal and business relations based on MLNs and a social network comprised of user profile data. In future work, a general method is needed for applying various social relations. In addition, an improvement in the ranking method is necessary.

References

1. Clements, M.: Personalization of social media, FDIA(BCS IRSG Symposium: Future Directions in Information Access (2007)
2. Choi, C., et al.: Travel ontology for intelligent recommendation system. In: Third Asia International Conference on Modelling & Simulation, AMS 2009. IEEE (2009)
3. Choi, C., et al.: Travel ontology for recommendation system based on semantic web. In: The 8th International Conference on Advanced Communication Technology, ICACT 2006, vol. 1. IEEE (2006)
4. http://en.wikipedia.org/wiki/Social_network_analysis
5. Choi, C., Choi, J., Kim, P.: Ontology based Access Control Model for Security Policy Reasoning in Cloud Computing. Journal of Supercomputing 67(3), 711–722 (2014)
6. Choi, C., Hwang, M., Choi, D., Choi, J., Kim, P.: Automatic Document Tagging using Online Knowledge Base. Information: An International Interdisciplinary Journal 14(5), 1709–1720 (2011)
7. Choi, J., Choi, C., Choi, D., Kim, J., Kim, P.: Automatic Extraction of Semantic Concept-Relation Triple Pattern from Wikipedia Articles. Information: An International Interdisciplinary Journal 15(7), 2755–2770 (2012)
8. http://research.microsoft.com/en-us/projects/socialinformationretrieval/
9. http://en.wikipedia.org/wiki/Social_search
10. Social Networks: Building Smart Communities through Network Weaving, Valdis Krebs (2002)
11. Choi, C., et al.: Probabilistic spatio-temporal inference for motion event understanding. Neurocomputing 122, 24–32 (2013)
12. de Oliveira, P.C.: Probabilistic Reasoning in the Semantic Web using Markov Logic. MSc Thesis (July 2009)

Computational Intelligence for Personalized Travel Scheduling System

Li-Fu Hsu[1], Chuin-Chieh Hsu[2], and Tsai-Duan Lin[3]

[1] Hwh Hsia Institute of Technology,
Department of Information Management, 235 Taipei, Taiwan
[2] National Taiwan University of Science and Technology,
Department of Information Management, 69042 Taipei, Taiwan
{Lhhsu,Lintd}@cc.hwh.edu.tw, CChsu @mail.ntust.edu.tw

Abstract. The important function of the Intelligent Transportation System (ITS) is to collect and disseminate certain information from different locations of the road network. The information includes traffic safety, current experienced travel times, or other information the travelers are interested in. The key point in optimizing a travel time lies at locating a shortest path under various predefined constraints or limits. Due to the constraints such as travel time limits, sightseeing-spot visit sequence and priority, and travel types, etc., the solutions are often difficult and time consuming to reach optimal. In this paper, we proposed Computational Intelligence for Personal Travel Scheduling Algorithm (CIPTS), which has two constraints, travel day and time limits. Using tree model to obtain the shortest paths and then merging the paths. CIPTS exhibits its effectiveness in resolving the personalized travel scheduling problems and compare with the Efficient Spot Tour(EST) method in our experiments. The CIPTS is promising to be a useful and realistic automatic tool for performing practical travel planning tasks. The efficiency of CIPTS improvement is even more significant when the number of sightseeing spots to be scheduled grows greater.

Keywords: Intelligent Transportation System, personalized travel scheduling, shortest path, sightseeing-spot, Dijkstra algorithm.

1   Introduction

A good travel planning tool can meet the emerging demands of independent travelers for easier travel itinerary arrangements in order to promote travel quality and a higher degree of satisfaction. Most of the travel agencies merely provide fixed itineraries and not meet the demand of personalized planning by independent travelers. Lacking effective travel planning tools, independent travelers have been forced to collect travel information from travel agencies or airline companies. The travelers should browse web pages of travel forums, and organize the information by themselves. The results are often far from optimal. This is an unwise solution when independent travelers plan their travel itinerary.

© Springer International Publishing Switzerland 2015
D. Camacho et al. (eds.), *New Trends in Computational Collective Intelligence*,
Studies in Computational Intelligence 572, DOI: 10.1007/978-3-319-10774-5_20

Consider the situation about travel scheduling as follows. An independent traveler is planning his/her family trip to a certain destination, e.g., Taiwan. Once he/she has collected necessary information about the area, decided the number of days to stay and the sightseeing spots to visit. The traveler may be interested in the history and cultural resources, special events, interesting festivals or scenery...etc. of destinations. Scheduling the visiting spots with minimum total travel time, within the period of staying may determine the quality of the trip most. To handle this kind of issue, we define the Computational Intelligence for personalized travel scheduling (CIPTS) problem as follows: given a starting nodes v0 as the accommodation place, a set of nodes S as visiting spots, and a pair of parameters $L = (l_d, l_h)$ as day limit in a trip and hour limit per day, respectively. The purpose is to find a tour includes all the nodes in S with the shortest travel time and satisfies the two limits. The limits here refer to how many days are planned for the trip and how many hours a day in maximum for sightseeing.

The process time to solve the CIPTS problem increases rapidly along with the expansion of the problem size. In this paper, we propose a new heuristic algorithm which combines scheduling approach in general optimization techniques to solve the PTS problem. The proposed computational intelligence for personal traveling scheduling algorithm (CIPTS) implement with the travel related constraints or personalized needs.

The rest of the article is organized as follows: in the next section, related works are presented. In section 3, we define the travel scheduling problem. Section 4 is proposed algorithm and an example for CIPTS algorithm. The article concludes with Section 5.

2   Related Work

The most similar CIPTS problem is the Traveling Salesman Problem (TSP), which schedules an itinerary path with the shortest distance that contains all the nodes exactly once. TSP is known as an NP-hard problem, and there exists a substantial literature for its solution [18]~[20].

When the solution of TSP is applied to other routing related applications, it starts further literature such as using genetic algorithms for transportation-route scheduling [2] and other scheduling problems [4],[5], the layout in a warehouse environment [6], travel scheduling with heuristics [3],[8], and Intelligent Transportation Systems [23].

The vehicle routing problem (VRP) is also similar to CIPTS problem. The VRP is a very complicated combinatorial optimization problem that has been worked on since the late fifties, because of its central meaning in distribution management. Problem specific methods [21], [22] as well as metaheuristic like tabu search, simulated annealing, genetic algorithms and neural networks have been proposed to solve it. The VRP and the TSP are closely related. As soon as the customer of the VRP is assigned to vehicles, the VRP is reduced to several TSPs.

However, the CIPTS problem differs from TSP and VRP. The TSP and the VRP are closely related. We only compare it with TSP in the following aspects. The first difference is that a vertex in a CIPTS graph possesses a value as its duration of staying time (properly in hours); but in TSP, the graph are assigned values to the edges. Secondly, CIPTS allows a vertex in its graph to be "passed" more than once (but usually stayed only once, the difference between pass through and stay will be stated in detail in the following sections), while TSP does not permit multiple pass through. Thirdly, CIPTS takes into consideration two limits: l_d as day limit in a trip and l_h as hour limit per day. Specifically, the daily schedule routes must not exceed l_d and the entire trip schedule must not exceed l_h. TSP traditionally considers only how to obtain the shortest path passing each node exactly once.

For CIPTS related problems, some researches adopt clustering analysis and sequential pattern mining techniques to select optimal sightseeing spots regarding urban paths, sightseeing paths, lodging and meals [3], [8]; some apply genetic algorithm in optimizing customer travel paths [18], and others such as the EST Algorithm [3] computes the start node and the shortest paths to other nodes in order to extract paths with smaller average costs.

The EST algorithm [3] finds all shortest paths from the starting node to the other selected nodes first. The second step is to select an efficient path, with the minimum average cost, from among them. The third step is to designate the last node in the efficient path to be a new starting node. The last step repeats the procedure until all selected nodes are included in the concatenated path.

3   Computational Intelligence for Personalized Travel Scheduling Problem

In CIPTS problem, a travel scheduling can be represented with an undirected graph $G = (V, E)$, where V is the set of sightseeing spots, $V = \{v_1, v_2, \ldots, v_n\}$ and $E \subseteq V \times V$, the set of any two sightseeing spots, edge $e_{ij} \in E$ represents the edge from sightseeing spot vi to another spot vj. The set S is represented by the sightseeing spots to be visited, and $S \subseteq V$. Weight $w(v_i)$ represents the visit time intended to be spent at sightseeing spot v_i and weight $w(e_{ij})$ denotes the travel time needed from spot v_i to spot v_j.

3.1   Assumptions

In solving the travel scheduling problem, we use some parameters which are defined in Table 1.

Table 1. Parameters used to formulate the travel scheduling problem

Parameters	Definition
V	The set of sightseeing spots
e_{ij}	The edge from sightseeing spot v_i to another spot v_j
S	The set of sightseeing spots to be visited
$w(v_i)$	The visit time intended to be spent at sightseeing spot v_i
$w(e_{ij})$	The traffic time needed from spot v_i to spot v_j
P	The set of the smallest paths
v_0	The accommodation place
$\varphi(p)$	The time cost of traffic and visit
l_d	day limit in a trip
l_h	hour limit per day

The assumptions of travel scheduling problem are made as follow:

(1) An independent traveler makes a selection of most interesting place to visit by
(a) history and culture (b) special events or interesting festivals (c) scenery and
landscape.
(2) The upper limit of the number of daily visit hours is assumed to be l_h hours
(excluding meal time), which is set by the traveler as a parameter.
(3) Visiting each sightseeing spot only once, but the same paths and sightseeing
spots can be passed through repeatedly when needed.
(4) An traveler begins his travel itinerary from the hotel each day, visits a portion
of the sightseeing spots and returns to the hotel to rest in the evening. He will
set off from the same hotel, again on the next day, until all the sightseeing spots
are visited.

Following the above assumptions, the proposed algorithm CIPTS extracts the most
interesting sightseeing spots from the graph into l_d tour paths each within l_h hours,
where l_d and l_h are defined previously as number of days in a trip and limited hours
per day.

3.2   Definitions

We first define the vertex v_0 corresponding to the accommodation place as the root
node which is defined as the only node in level 1 in the travel graph.

As mentioned previously, $w(v_i)$ represents the sightseeing time length at spot v_i and
$w(e_{ij})$ is the traffic cost at edge e_{ij}. Let $\varphi(p)$ denote the cost (both traffic and visit) at
sightseeing spot path. The shortest traveling path from v0 throughout sightseeing
spots v_i is minimized and the formula is $\varphi(p) = \Sigma (w(v_i) + w(e_k))$, where
$v_i \in \Sigma$, $e_k \in \Pi$.

Like a traveling salesman problem (TSP), the CIPTS problem is more difficult than
TSP. The number of possible efficient tours usually grows exponentially with the
problem size and finding them is difficult. The objective of CIPTS problem is to
minimize the time traveled and meet the traveler's requirements in time limits and
sightseeing spots.

Once l_d and l_h are decided, the traveler needs to select the sightseeing spots to visit.
We also let $|P|$ be the number of the paths having the smallest cost, and $|S|$ the number
of sightseeing spots the traveler wants to visit. In the CIPTS problem, we first have G
$= (V, E)$, where V is the set of sightseeing spots, $V = \{v_1,v_2,.....,v_n\}$, $|V|=n$; E is the set
of all edges. A set S is a node set to be selected by a traveler from the set of V. The
goal of the proposed CIPTS algorithm is to accomplish a minimized daily traveling
time by optimal travel scheduling. Before going further, we need some definitions
listed below.

Definition 1. A path $p \in P$ is said be efficient if and only if there does not exist $p' \in P$
such that $\varphi_i(p') <= \varphi_i(p)$, the cost at sightseeing spot v_i, where $i=1, 2,....., n$, and $\varphi_i(p')$
$\neq \varphi_i(p)$ for at least one i. If there exists such a p', p is said to be inefficient.

Definition 2. If p is efficient, then $\varphi(p)$ is said to be nondominated; if p is inefficient, $\varphi(p)$ is said to be dominated.

Definition 3. A path $p \in P$ is said the shortest that has minimum cost. The cost is including travel time and visited time.

The task of CIPTS is to identify a set of sightseeing spots to be visited, S, in determining a provisional schedule and a smallest path set, P, which is empty initially. To resolve the CIPTS problem, we want to find the shortest path from the root v_0 throughout the destination nodes. If a shortest path found is greater than l_h, the path will be discarded. The time spent at each sightseeing spot is defined as

$$w(v_i) = \begin{cases} 0 & if \quad v_i \quad is \quad not \quad visited \\ w(v_i) & if \quad v_i \quad is \quad visitid \end{cases} \tag{1}$$

The shortest path from v_0 to a destination node v_i is expressed as

$$\varphi(p) = \min\{ \sum_{v_i \in S, e_k \in P} (w(v_i) + w(e_k)) \} \tag{2}$$

for $i = 1, 2, \ldots, n$, and $j = 1, 2, \ldots, n$.

To obtain the shortest paths for the sightseeing spots, we need to determine the spots, with limited sightseeing hours l_h assigned to a day and other constraints to be complied with, such that the shortest paths are minimized with all given constraints satisfied. Further theorem needed for the proposed CIPTS is derived next.

Theorem I. An efficient travel itinerary does not contain any inefficient paths.

Proof

Consider a PTS problem having a node set $V = \{v_1, v_2, \ldots, v_n\}$ and let p'_{ij}, $p'_{ij} \in P$, be an inefficient path between Nodes v_i and v_j. Then, there must exist at least one p''_{ij}, $p''_{ij} \in P$, such that the following inequalities hold.

$$\varphi_i(p''_{ij}) \leq \varphi_i(p'_{ij}) \quad \text{for } i = 1, 2, \ldots, n \tag{3}$$

$$\varphi_i(p''_{ij}) < \varphi_i(p'_{ij}) \quad \text{for at least one } i \tag{4}$$

Letting a set, A_7, contain Nodes, (v_a, v_b), such that an inefficient path, p'_{ab}, between Nodes v_a and v_b is included in the set of the smallest path P. The cost of the smallest path P in objective k can be expressed by

$$\varphi_k(P) = \sum_{(v_a,v_b) \in A\tau - (v_i,v_j)} \varphi_k(p'_{ab}) + \varphi_k(p'_{ij}) \tag{5}$$

for $i=1,2,\ldots,n$

Replacing p'_{ab} in the set of the smallest path P with p''_{ab}, we obtain a t the set of the smallest path P' having a   cost in objective k such that

$$\varphi_k(P') = \sum_{(v_a,v_b) \in A\tau - (v_i,v_j)} \varphi_k(p'_{ab}) + \varphi_k(p'_{ij}) \tag{6}$$

for $i=1,2,\ldots,n$

Substituting Eqs. (3) and (4) into $\varphi_k(P)$ and $\varphi_k(P')$, we arrive at

$$\varphi_k(p') \le \varphi_k(p) \qquad \text{for } k=1, 2,\ldots, n$$

$$\varphi_k(p') < \varphi_k(p) \qquad \text{for at least one } k$$

Therefore, P cannot be an efficient smallest path. Q.E.D.

Similarly, a dominated sub-travel itinerary cannot be part of an efficient travel itinerary even if it is made up of efficient paths.

4   The Proposed Algorithm CIPTS

According to the definitions and theorem, the Computational Intelligence for Personal Travel Scheduling (CIPTS) Algorithm is proposed. The pseudo code is shown in Figure 1.

The algorithm of PTS

```
1    Input : Graph, day limit in a trip ld hour limit per day lh
2              The accommodation place v0, the set of visited spots S
3    Output : The set of the smallest paths P
4    begin
5          Tree = Tree_mode(Graph, v0);
6          P' = Shortest_path(Tree, S, ld , lh );
7          // P' is the set of the shortest paths
8          P = Shortest_path(Tree, S, ld , lh );
9          // P is the set of the shortest paths after merging
10   end
```

Fig. 1. Pseudo code of the PTS algorithm

4.1   The Shortest Paths Stage

The shortest paths are obtain by improved Dijkstra Algorithm to executed with the following steps.

Step 1: Let set S be the node of sightseeing spots to be visited. Set P is the traveling paths, which is empty initially.

Step 2: Use the Eq. (2), $\varphi(\ ) = \min \Sigma(\ \ ) + \ \ ))$, where $v_i \in S$, $e_k \in P$,to find the shortest path (with minimum cost) from the root to a destination node, v_i. Assuming m shortest paths are found, add them to set P.

Step 3: Remove all paths having a cost greater than l_h, the number of daily traveling hours predetermined by the traveler.

Step 4: Remove node v_i from set S, if the shortest path of which has been included in set P.

Step 5: Go to Step 2 until set S is empty.

4.2   The Paths Merge Stage

We will merge the shortest paths with the following steps and criteria:

Step 1: Letting the number of traveling days the traveler predetermined be l_d, which determine if $l_d > |P|$, the number of the paths having the smallest cost. If yes, the result indicates that the traveler wants a relaxed traveling scheduling, PTS will schedule at least one or more days for travel itinerary one path. If no, go to Step 2.

Step 2: Verify if $l_d = |P|$. If yes, PTS will schedule the travel itinerary for one (different) path per day. If no, go to Step 3.

Step 3: Verify if $l_d < |P|$. If yes, PTS will merge the shortest paths to one path, while limiting the number of daily traveling hours to less than or equal to l_h, until $l_d \geq |P|$. Then, stop. If making $l_d \geq |P|$ is not possible, stop merging and halt.

During the process of combining shortest paths, we shall repeatedly go back the top vertex, v_0, and the time of repeated travels through v_0 shall be all counted into the travel time. Hence, the paths with shorter time shall take a higher priority to be combined first. Unless there exist travel sequence constraints and the constraints can be met, the paths with longer time shall be ignored, and not put into calculation to save execution time.

When many shortest paths meet the combining requirements, all the resultant combined paths must be re-examined to see if the sightseeing spots with shorter costs belong to priority visit targets and whether the time required for the paths meet the daily travel itinerary time constraint or not. Otherwise, the combined paths must be discarded.

4.3   An Example for CIPTS Algorithm

An example is shown in Figure 2 and Figure 3, $V=\{A,B,C,D,E,F,G,H,I,J\}$, and V_0 is node G, S=$\{A, D, E, F, H, I, J\}$. The converted result by the PTSA is shown in Figure 8. Assuming that day limit in a trip is four days (l_d =4) and that hour limit per day is set to eight hours (l_h=8), the results are listed as follows. In order to describe clearly, we make the tree-type graph to represent the shortest paths from root to the sightseeing nodes which traveler wants to visit. The visited nodes are represented by yellow nodes and not visited nodes are used by white nodes. The results are listed as follows:

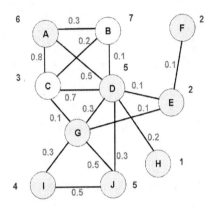

Fig. 2. An example for travel graph

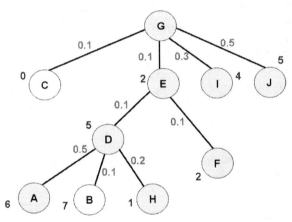

Fig. 3. A tree corresponding to Fig. 2 produced by PTSA

After intelligent computing, the visit paths are shown in Table 2.

Table 2. The paths and their costs

Path	Cost
Path $G{\rightarrow}E{\rightarrow}D$	$\varphi\,(GED)$= 5+0.1+2+0.1+=7.2
Path $G \xrightarrow{ED} A$	$\varphi\,(G_{ED}A)$= 6+0.1+ 0.1+0.5=6.7
Path $G \xrightarrow{ED} H$	$\varphi\,(G_{ED}H)$= 1+ 0.1+ 0.1+0.2=1.4
Path $G \xrightarrow{E} F$	$\varphi\,(G_{E}F)$= 2+ 0.1+ 0.1=2.2
Path $G{\rightarrow}I$	$\varphi\,(GI)$= 4+0.3=4.3
Path $G{\rightarrow}J$	$\varphi\,(GJ)$= 5+0.5=5.5

The final suggestion chooses path P', $P' = \{GED, G_{ED}A, G_EF_{EG}J, GI_{GED}H \}$, for this path requires a total cost of only 13.9 hours, slightly smaller than that of the other path, $P = \{ GED, G_{ED}A, G_EF_{EG}I, G_{ED}H_{DEG}J \}$ which requires 14.0 hours. In other words, path P' has the greatest travel efficiency among the paths created. In this study, the example is confirmed to be simple and effective in CIPTS after numerous and repeated applications. The sample example is shown above, with a superior execution time performance to be described next.

5 Conclusion and Prospective Outlook

This paper presents a new heuristic algorithm to schedule the sightseeing spots for travelers. The Computational Intelligence for personalized travel scheduling (CIPTS) problem is more difficult than TSP and VRP since added constraints by related parameters, such as day and time limits. In this study, we propose a personalized solution for CIPTS problem which will not only simplify travel graphs into tree diagram for easier analysis and inspection but also create an optimal itinerary in compliance with the mentioned constraints.

The performance of CIPTS has been observed by its application on some practical graphs for different experiment data sets. The CIPTS algorithm is promising to be a useful and realistic automatic tool for performing practical travel planning tasks.

References

1. Hsu, L.F., Hsu, C.C., Lin, T.D.: Data Mining in Personalized Travel Information System. In: The 2nd International Conference on Information Technology Convergence and Services, pp. 2–5 (2010)
2. Liu, C.-H.: Using genetic algorithms for the coordinated scheduling problem of a batching machine and two-stage transportation. Applied Mathematics and Computation 217(24), 10095–10104 (2011)
3. Huang, Y.-F., Chen, C.-N.: Implementation for the Arrangement and mining analysis of Traveling Schedules. Journal of Information Science and Engineering 22, 23–146 (2006)
4. Amirthagadeswaran, K., Arunachalam, V.: Enhancement of performance of genetic algorithm for job shop scheduling problems through inversion operator. International Journal of Advanced Manufacturing Technology 32(7/8), 780–786 (2007)
5. Naso, D., et al.: Genetic algorithms for supply-chain scheduling: A case study in the distribution of ready-mixed concrete. European Journal of Operational Research 177(3), 2069–2099 (2007)
6. Zhang, C., Yao, X., Yang, J.: An evolutionary approach to materialized views selection in a data warehouse environment. IEEE Transactions On Systems Man and Cybernetics Part C-Application and Reviews 31(3), 282–294 (2001)
7. Biethahn, J., Nissen, V.: Evolutionary Algorithms in Management Applications. Springer, Berlin (1995)
8. Christoph, H., Stefan, I.: Vehicle Routing Problems with Inter-Tour Resource Constraints. Operating Research, 421–444 (2008)
9. Louis, S.J., Rawlins, G.J.: Designer genetic algorithm in structure design. In: Proc. 4th Int. Conf. Genetic Algorithms, pp. 53–60 (1991)

10. Syswerda, G.: Uniform crossover in genetic algorithms. In: Proc. 3rd Int. Conf. Genetic Algorithms, pp. 2–9 (1989)
11. Zheng, J.Y., Shi, M.: Mapping cityscapes to cyber space. In: Proceedings of International Conference on Cyberworlds, pp. 166–173 (2003)
12. Cormen, T.H., Leiserson, C.E., Rivest, R.L.: Introduction to Algorithms. MIT Press and McGraw-Hill (1990)
13. Kruskal, J.B.: On the shortest spanning subtree of a graph and the traveling salesman problem. Proceedings of the American Mathematical Society, 48–50 (February 1956)
14. Chu, Y.J., Liu, T.H.: On the shortest arborescence of a directed graph. Science Sinica 14, 1396–1400 (1965)
15. Edmonds, J.: Optimum branchings. J. Research of the National Bureau of Standards, 233–240 (1967)
16. Sauer, J.G., et al.: A discrete differential evolution approach with local search for traveling salesman problems. Studies in Computational Intelligence 357, 1–12 (2011)
17. Tezcaner, D., Köksalan, M.: An Interactive Algorithm for Multi-objective Route Planning. Journal of Optimization Theory and Applications 150(2), 379–394 (2011)
18. Kalmár-Nagy, T., Giardini, G.: Genetic algorithm for combinatorial path planning: The subtour problem. Mathematical Problems in Engineering (2011)
19. Geng, X., et al.: Solving the traveling salesman problem based on an adaptive simulated annealing algorithm with greedy search. Applied Soft Computing Journal 11(4), 3680–3689 (2011)
20. Manthey, B.: Deterministic algorithms for multi-criteria TSP. In: Ogihara, M., Tarui, J. (eds.) TAMC 2011. LNCS, vol. 6648, pp. 264–275. Springer, Heidelberg (2011)
21. Clarke, G., Wright, J.W.: Scheduling of Vehicles from a Central Depot to a Number of Delivery Points. Oper. Res., 568–581 (1964)
22. Gillett, B.E., Miller, L.R.: A Heuristic Algorithm for the Vehicle Dispatch Problem. Oper. Res., 340–347 (1974)
23. Laborczi, P., Mezny, B., Török, A., Zoltan, R.: Query-based Information Gathering in Intelligent Transportation Systems. Electronic Notes in Discrete Mathematics 36, 1201–1208 (2010)
24. Dantzig, G.B., Ramser, J.H.: The truck dispatching problem. Management Science 6(1), 80–91 (1959)

Author Index

Printed in the United States
By Bookmasters